Avatars of Wizardry

Avatars of Wizardry

Poetry Inspired by
George Sterling's "A Wine of Wizardry" and
Clark Ashton Smith's "The Hashish-Eater"

With a Foreword by S. T. Joshi
Compiled and Edited by Charles Lovecraft

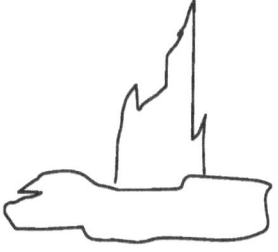

P'REA PRESS

Sydney, Australia
2012

P'rea Press, Sydney, Australia, www.preapress.com
First edition, November 2012; reprint February 2016.
All rights reserved. Reviewers may quote short passages.

Copyright © by P'rea Press 2012, 2016.• Cover art © by Gavin L. O'Keefe 2012, 2016.• Frontispiece © by David Schembri Studios 2012, 2016.• Foreword © by S. T. Joshi 2012, 2016.• Publisher's Preface © by Charles Lovecraft 2012, 2016.• A Wine of Wizardry © by the Bancroft Library 2012, 2016.• The Hashish-Eater; or, The Apocalypse of Evil © by CASiana Enterprises 2012, 2016.• Visions of Golconda © by Richard L. Tierney 2012, 2016.• Memoria: A Fragment from the Book of Wyvern © by Leigh Blackmore 2012, 2016.• A Trip to the Hypnotist © by Alan Gullette 2012, 2016.• Thirteen Ways of Looking At and Through Hashish © by Bruce Boston 2012, 2016.• The Mantle of Merlin © by Earl Livings 2012, 2016.• The Necromantic Wine © by Wade German 2012, 2016.• Sandalwood © by Michael Fantina 2012, 2016.• Lucubration © by Kyla Lee Ward 2012, 2016.

"A Wine of Wizardry" published by permission of the Bancroft Library, University of California, Berkeley, representing the Estate of George Sterling. • "A Trip to the Hypnotist" published by permission of Hippocampus Press, New York. • "The Hashish-Eater" published by permission of CASiana Enterprises, representing the Estate of Clark Ashton Smith.

Cover created and designed by Gavin L. O'Keefe. • Frontispiece designed by David Schembri Studios. • Publisher's logo designed by Charles Lovecraft. • Book designed by David E. Schultz. • Set in Goudy Old Style type 11 point. • Printed by Lightning Source International.

A sincere thank you to all.

National Library of Australia Cataloguing-in-Publication entry:
Title: Avatars of wizardry : poetry inspired by George Sterling's "A wine of wizardry" and Clark Ashton Smith's "The hashish-eater" / George Sterling, Clark Ashton Smith; foreword, S. T. Joshi; editor, Charles Lovecraft.
ISBN: 9780980462586 (pbk)
Subjects: Imagist poetry.
            Narrative poetry.
            Fantasy poetry.
            Supernatural in literature.
Other Authors/Contributors:
            Smith, Clark Ashton, 1893–1961. The hashish-eater.
            Sterling, George, 1869–1926. A wine of wizardry.
            Joshi, S. T., 1958–
            Lovecraft, Charles.
Dewey Number: 808.81

For George and Clark

Contents

Foreword, *by* S. T. Joshi	9
Publisher's Preface	13
A Wine of Wizardry *George Sterling*	19
The Hashish-Eater; or, The Apocalypse of Evil *Clark Ashton Smith*	27
Visions of Golconda *Richard L. Tierney*	45
Memoria: A Fragment from the Book of Wyvern *Leigh Blackmore*	49
A Trip to the Hypnotist *Alan Gullette*	55
Thirteen Ways of Looking At and Through Hashish *Bruce Boston*	59
The Mantle of Merlin *Earl Livings*	77
The Necromantic Wine *Wade German*	83
Sandalwood *Michael Fantina*	93
Lucubration *Kyla Lee Ward*	99
Select Bibliography	107
About the Contributors	109

FOREWORD

It is widely acknowledged that, for all the excellence of August Derleth's landmark compilation, *Dark of the Moon: Poems of Fantasy and the Macabre* (1947), one of its major omissions was the work of George Sterling, notably his imperishable 210-line poem "A Wine of Wizardry" (1904). Perhaps Derleth omitted it because of its length, although he seems to have felt no compunction in including the nearly 600 lines of "The Hashish-Eater" (1920) by his friend Clark Ashton Smith, along with numerous other poems by the California poet. The omission of Sterling is doubly odd, because Derleth received substantial assistance in the assembling of *Dark of the Moon* from Donald Wandrei, whose sensitivity to weird poetry was much greater than Derleth's own and who was certainly familiar with Sterling's variegated work.

In any event, "A Wine of Wizardry" and "The Hashish-Eater" have become the twin pillars of weird poetry in the twentieth century, equalled perhaps only by Wandrei's *Sonnets of the Midnight Hours* (1927) and H. P. Lovecraft's *Fungi from Yuggoth* (1929–30). And just as Lovecraft's sonnet cycle has inspired any number of brilliant imitations—including, most recently, Ann K. Schwader's *In the Yaddith Time* (2007) and Leigh Blackmore's *Spores from Sharnoth* (2008)—the Sterling and Smith epics have, as this volume triumphantly shows, inspired some of our leading weird poets to use them as springboards for fantastic visionings of their own.

When, in early 1904, Sterling quoted the two most famous lines of "A Wine of Wizardry" ("The blue-eyed vampire, sated at her feast, / Smiles bloodily against the leprous moon") in a letter to Ambrose Bierce, the latter told him that the lines "give me the shivers. Gee! they're awful!" Bierce undertook a three-year effort to secure publication of the poem, and it finally appeared in *Cosmopolitan*

(September 1907), with a laudatory article by Bierce, "A Poet and His Poem." The flamboyant imagery of the poem, not to mention Bierce's perhaps extravagant praise, led many to scoff at the poem and at Bierce's critical judgment; but Bierce shot them down in a pungent rebuttal, "An Insurrection of the Peasantry." One comment deserves particular attention. In regard to those famous lines about the "blue-eyed vampire," one critic opined: "Somehow one does not associate blue eyes with a vampire." Bierce replied tartly: "Of course it did not occur to him that this was doubtless the very reason why the author chose the epithet—if he thought of anybody's conception but his own. 'Blue-eyed' connotes beauty and gentleness; the picture is that of a lovely, fair-haired woman with the telltale blood about her lips. Nothing could be less horrible; nothing more terrible."

Just as Bierce had acted as a poetic mentor to Sterling, so did Sterling fill that same function when he came into contact with the boy prodigy from Auburn, Clark Ashton Smith. It was Sterling who shepherded Smith's early volumes of poetry—*The Star-Treader and Other Poems* (1912), *Odes and Sonnets* (1918), *Ebony and Crystal* (1922), and *Sandalwood* (1925)—into print, making sure they were widely reviewed (in California, at any rate) and even reviewing one of them himself under a pseudonym.

When Smith, in early 1920, told Sterling that he was working on "The Hashish-Eater" Sterling wrote back enthusiastically ("I'm greatly interested in the haschich poem"), and expressed even greater approbation when he read the poem:

> "The Hashish-Eater" is indeed an amazing production. My friends will have none of it, claiming it reads like an extension of "A Wine of Wizardry." But I think there are many differences, and at any rate, it has more imagination in it than in any other poem I know of. Like the "Wine," it fails on the esthetic side, a thing that seems of small consequence in a poem of that nature.

The opinions of Sterling's "friends" seem to suggest a lack of critical judgment or poetic imagination when dealing with fantastic poetry, since the two poems really have few similarities aside from their bizarre imagery. But the personal and aesthetic relations between the two poets will perennially link these epics of the imagination, demonstrating (as the poetry of Wandrei, Lovecraft, Frank Belknap Long, Robert E. Howard, Walter de la Mare, and many others also attests) that the early decades of the twentieth century constituted something of a golden age in weird verse.

But we seem to be in the midst of another renaissance of fantastic poetry, as the present volume attests. Richard L. Tierney, Leigh Blackmore, Wade German, Michael Fantina, and the other poets in this book have done far more than write mechanical pastiches of Smith's and Sterling's poems; they have found in their work the inspiration to weave a tapestry of weirdness that stands on its own as a substantial contribution to the fantastic verse of our own time. Poetry in general, and weird poetry in particular, may never have the wide audience of the routine bestselling novel, but connoisseurs know what aesthetic pleasures are in store for them when they read vivid, meticulously crafted work such as is contained in this book.

—S. T. JOSHI

Seattle, Washington
August 2012

Publisher's Preface

"Dreams of stained twilights of the South . . ."
—GEORGE STERLING

Anything is Possible in a Fantasy Imaginarium

*A*vatars of Wizardry was conceived serendipitously from George Sterling's creative influence upon Clark Ashton Smith and subsequently Smith's own creative influence upon Richard L. Tierney, and Leigh Blackmore. When, in 2008, Blackmore told me he had been inspired by Smith's "The Hashish-Eater" to write his poem "Memoria" (1976) and, in 2009, Tierney sent me his "Visions of Golconda" (1959), humbly stating, "It was my feeble attempt to follow the style of CAS's 'The Hashish Eater,'" I realised there was a spectacular conceptual continuity afoot. Independently, Tierney's and Blackmore's poems had sprung from the same spring, the flowing waters of a darker Pierian Spring, and now formed, to all intents and purposes, a unique trilogy with Smith's piece that spanned nearly a century.

Later, in 2010, as I read letters between Sterling and Smith collected in *The Shadow of the Unattained* (Hippocampus Press, 2005), I suddenly became aware there was a still earlier influence upon "Hashish-Eater." This was Sterling's intoxicating masterpiece, "A Wine of Wizardry," published in 1907, and which had created for Smith at the tender age of fifteen years "one of the defining moments of his youthful aesthetic life" (Schultz/Joshi, *Unattained*, 9). In a letter of Smith's to Sterling, dated 10 July 1920, Smith notes the influence of "A Wine of Wizardry" and Gustave Flaubert's *The Temptation of St. Anthony* upon "The Hashish-Eater," stating, "The 'Wine of Wizardry' has always seemed the ideal poem to me" (*Unattained*, 184), elsewhere calling it "the most colorful, exotic, and, in places, macabre, of Sterling's poems" (*Unattained*, 296). This deepened my resolve. I *had* to bring this book into existence.

Inspired by the splendid creative link between poets of different times, generations, and places, I set about obtaining copyright permission to publish Sterling's and Smith's two beacons of fantastic poetry and seeking contributions from contemporary poets and authors. Thus the last six poems were specially commissioned for this anthology and, except for "A Visit to the Hypnotist," they are published here for the very first time.

Sterling's trail-blazing "Wine of Wizardry" appears first in this volume, with the poems of Smith, Tierney, and Blackmore hotly pursuing as spontaneous avatars of Sterling's mystery and imagination. The balance of the poems are contemporary 21st century responses to Sterling's and Smith's turbulent "streams of violet midnight" inspiration, poured deluge-like through luminous realms of intellect, the subconscious, passion, fantasy, and other spaces. Rising spectacularly in answer are America's notable fantasy and science fiction poets Bruce Boston, Michael Fantina, Alan Gullette, Wade German, and Australia's Earl Livings and Kyla Lee Ward. Readers will find much in our contemporaries' work to prove the truth, "Anything is possible in a fantasy imaginarium."

This collection has further offerings. It celebrates the traditional poetic form of English-language narrative blank verse. Eight of the ten poems found here are in narrative blank verse form, a form greatly admired and used also by Shakespeare, Milton, Shelley, Keats, Cowper, Wordsworth, and Tennyson—giants all in the English-language literary tradition. *Avatars of Wizardry* carries on this tradition brightly and well, in the luminaries presented within its pages.

Now although "A Wine of Wizardry" is not blank verse per se in the fact it rhymes, nevertheless the poem's iambic pentameter powerfully reflects classic blank-verse sweep, rhythm and narrative structure, in the weaving and interlacing of long lustrous sentences, over and through the subsequent shining, wickerwork-like lines of the poem, in the very best traditions of blank verse enjambment-style poetry. Boston's wonderful 510-line epic, "Thirteen Ways of Looking At and Through Hashish," is exceptional in its employment of an improvised free verse mood fusion that retains "a definite influence of mood/ flavor/imagery from Smith (and from Wallace Stevens . . .)" (Boston email, 20 October 2012).

    Great poetry and great poems will always outlive the eras in which they are composed. It is the way of humans to remember greatness. Thus this collection celebrates more than a century of enthralling influence by "A Wine of Wizardry," now 108 years old. In ruddy-litten treasure-crypts out of time they might be reciting its lines at this very moment and thrilling at the cadences its "crimson banners" flutter and caress like fingers on the air. In similar fashion, its slightly younger contemporary, "The Hashish-Eater" is nearing its own centenary in 2020 and continues to thrill, with its own mind-expanding tendencies, the readers who come near. Together they hearken back to the post-Victorian era in which their two creators thrived as friends and poets. They continue to inspire rich fantasies among poets today, as the following two examples will attest. In German's poem, for instance, "The inspirational cocktail for 'The Necromantic Wine' was one part Sterling's 'Wine . . .,' one part Smith's 'Hashish- . . .'" (German email, 21 October 2012); and the inspiration behind Ward's "Lucubration" "was indeed 'A Wine of Wizardry,'" as Ward conveys (in email, 21 October 2012).

    Avatars of Wizardry has come about from the deep-seated love and rich regard for the English-language poetry tradition that I have gained through a lifetime of reading. P'rea Press is proud to proffer its fragrance to you. I can only wish you the most happy of sojourns as you meander through the weird poetry flowers here composed and grown just for you, and daisy chain them round your heart and mind. From such enchanted scented chains as these are links of beauty forged forever, and that could never pass into nothingness.

                                     —CHARLES LOVECRAFT, P'REA PRESS, 2012

Avatars of Wizardry

A Wine of Wizardry

George Sterling

When mountains were stained as with wine
By the dawning of Time, and as wine
Were the seas.
                    —AMBROSE BIERCE

Without, the battlements of sunset shine,
'Mid domes the sea-winds rear and overwhelm.
Into a crystal cup the dusky wine
I pour, and, musing at so rich a shrine,
I watch the star that haunts its ruddy gloom.
Now Fancy, empress of a purpled realm,
Awakes with brow caressed by poppy-bloom,
And wings in sudden dalliance her flight
To strands where opals of the shattered light
Gleam in the wind-strewn foam, and maidens flee
A little past the striving billows' reach,
Or seek the russet mosses of the sea,
And wrinkled shells that lure along the beach,
And please the heart of Fancy; yet she turns,
Tho' trembling, to a grotto rosy-sparred,
Where wattled dragons redly gape, that guard
A cowled magician peering on the damned
Thro' vials wherein a splendid poison burns,
Sifting Satanic gules athwart his brow.
So Fancy will not gaze with him, and now
She wanders to an iceberg oriflammed
With rayed, auroral guidons of the North—

Wherein hath winter hidden ardent gems
And treasuries of frozen anadems,
Alight with timid sapphires of the snow.
But she would dream of warmer gems, and so
Ere long her eyes in fastnesses look forth
O'er blue profounds mysterious whence glow
The coals of Tartarus on the moonless air,
As Titans plan to storm Olympus' throne,
'Mid pulse of dungeoned forges down the stunned,
Undominated firmament, and glare
Of Cyclopean furnaces unsunned.

Then hastens she in refuge to a lone,
Immortal garden of the eastern hours,
Where Dawn upon a pansy's breast hath laid
A single tear, and whence the wind hath flown
And left a silence. Far on shadowy tow'rs
Droop blazoned banners, and the woodland shade,
With leafy flames and dyes autumnal hung,
Makes beautiful the sunset of the year.
For this the fays will dance, for elfin cheer,
Within a dell where some mad girl hath flung
A bracelet that the painted lizards fear—
Red pyres of muffled light! Yet Fancy spurns
The revel, and to eastern hazard turns,
And glaring beacons of the Soldan's shores,
When in a Syrian treasure-house she pours,
From caskets rich and amethystine urns,
Dull fires of dusty jewels that have bound
The brows of naked Ashtaroth around.
Or hushed, at fall of some disastrous night,
When sunset, like a crimson throat to hell,
Is cavernous, she marks the seaward flight
Of homing dragons dark upon the West;
Till, drawn by tales the winds of ocean tell,
And mute amid the splendors of her quest,

To some red city of the Djinns she flees
And, lost in palaces of silence, sees
Within a porphyry crypt the murderous light
Of garnet-crusted lamps whereunder sit
Perturbéd men that tremble at a sound,
And ponder words on ghastly vellum writ,
In vipers' blood, to whispers from the night—
Infernal rubrics, sung to Satan's might,
Or chaunted to the Dragon in his gyre.
But she would blot from memory the sight,
And seeks a stainéd twilight of the South,
Where crafty gnomes with scarlet eyes conspire
To quench Aldebaran's affronting fire,
Low sparkling just beyond their cavern's mouth,
Above a wicked queen's unhallowed tomb.
There lichens brown, incredulous of fame,
Whisper to veinéd flowers her body's shame,
'Mid stillness of all pageantries of bloom.
Within, lurk orbs that graven monsters clasp;
Red-embered rubies smolder in the gloom,
Betrayed by lamps that nurse a sullen flame,
And livid roots writhe in the marble's grasp,
As morning airs invoke the conquered rust
Of lordly helms made equal in the dust.
Without, where baleful cypresses make rich
The bleeding sun's phantasmagoric gules,
Are fungus-tapers of the twilight witch
(Seen by the bat above unfathomed pools)
And tiger-lilies known to silent ghouls,
Whose king hath digged a somber carcanet
And necklaces with fevered opals set.
But Fancy, well affrighted at his gaze,
Flies to a violet headland of the West,
About whose base the sun-lashed billows blaze,
Ending in precious foam their fatal quest,
As far below the deep-hued ocean molds,

With waters' toil and polished pebbles' fret,
The tiny twilight in the jacinth set,
The wintry orb the moonstone-crystal holds,
Snapt coral twigs and winy agates wet,
Translucencies of jasper, and the folds
Of banded onyx, and vermilion breast
Of cinnabar. Anear on orange sands,
With prows of bronze the sea-stained galleys rest,
And swarthy mariners from alien strands
Stare at the red horizon, for their eyes
Behold a beacon burn on evening skies,
As fed with sanguine oils at touch of night.
Forth from that pharos-flame a radiance flies,
To spill in vinous gleams on ruddy decks;
And overside, when leap the startled waves
And crimson bubbles rise from battle-wrecks,
Unresting hydras wrought of bloody light
Dip to the ocean's phosphorescent caves.

So Fancy's carvel seeks an isle afar,
Led by the Scorpion's rubescent star,
Until in templed zones she smiles to see
Black incense glow, and scarlet-bellied snakes
Sway to the tawny flutes of sorcery.
There priestesses in purple robes hold each
A sultry garnet to the sea-linkt sun,
Or, just before the colored morning shakes
A splendor on the ruby-sanded beach,
Cry unto Betelgeuse a mystic word.
But Fancy, amorous of evening, takes
Her flight to groves whence lustrous rivers run,
Thro' hyacinth, a minster wall to gird,
Where, in the hushed cathedral's jeweled gloom,
Ere Faith return, and azure censers fume,
She kneels, in solemn quietude, to mark
The suppliant day from gorgeous oriels float

And altar-lamps immure the deathless spark;
Till, all her dreams made rich with fervent hues,
She goes to watch, beside a lurid moat,
The kingdoms of the afterglow suffuse
A sentinel mountain stationed toward the night—
Whose broken tombs betray their ghastly trust,
Till bloodshot gems stare up like eyes of lust.
And now she knows, at agate portals bright,
How Circe and her poisons have a home,
Carved in one ruby that a Titan lost,
Where icy philtres brim with scarlet foam,
'Mid hiss of oils in burnished caldrons tost,
While thickly from her prey his life-tide drips,
In turbid dyes that tinge her torture-dome;
As craftily she gleans her deadly dews,
With gyving spells not Pluto's queen can use,
Or listens to her victim's moan, and sips
Her darkest wine, and smiles with wicked lips.
Nor comes a god with any power to break
The red alembics whence her gleaming broths
Obscenely fume, as asp or adder froths,
To lethal mists whose writhing vapors make
Dim augury, till shapes of men that were
Point, weeping, at tremendous dooms to be,
When pillared pomps and thrones supreme shall stir,
Unstable as the foam-dreams of the sea.

But Fancy still is fugitive, and turns
To caverns where a demon altar burns,
And Satan, yawning on his brazen seat,
Fondles a screaming thing his fiends have flayed,
Ere Lilith come his indolence to greet,
Who leads from hell his whitest queens, arrayed
In chains so heated at their master's fire
That one new-damned had thought their bright attire
Indeed were coral, till the dazzling dance

So terribly that brilliance shall enhance.
But Fancy is unsatisfied, and soon
She seeks the silence of a vaster night,
Where powers of wizardry, with faltering sight
(Whenas the hours creep farthest from the noon)
Seek by the glow-worm's lantern cold and dull
A crimson spider hidden in a skull,
Or search for mottled vines with berries white,
Where waters mutter to the gibbous moon.
There, clothed in cerements of malignant light,
A sick enchantress scans the dark to curse,
Beside a caldron vext with harlots' blood,
The stars of that red Sign which spells her doom.

Then Fancy cleaves the palmy skies adverse
To sunset barriers. By the Ganges' flood
She sees, in her dim temple, Siva loom
And, visioned with a monstrous ruby, glare
On distant twilight where the burning-ghaut
Is lit with glowering pyres that seem the eyes
Of her abhorrent dragon-worms that bear
The pestilence, by Death in darkness wrought.
So Fancy's wings forsake the Asian skies,
And now her heart is curious of halls
In which dead Merlin's prowling ape hath spilt
A vial squat whose scarlet venom crawls
To ciphers bright and terrible, that tell
The sins of demons and the encharneled guilt
That breathes a phantom at whose cry the owl,
Malignly mute above the midnight well,
Is dolorous, and Hecate lifts her cowl
To mutter swift a minatory rune;
And, ere the tomb-thrown echoings have ceased,
The blue-eyed vampire, sated at her feast,
Smiles bloodily against the leprous moon.

But evening now is come, and Fancy folds
Her splendid plumes, nor any longer holds
Adventurous quest o'er stainéd lands and seas—
Fled to a star above the sunset lees,
O'er onyx waters stilled by gorgeous oils
That toward the twilight reach emblazoned coils.
And I, albeit Merlin-sage hath said,
"A vyper lurketh in ye wine-cuppe redde,"
Gaze pensively upon the way she went,
Drink at her font, and smile as one content.

THE HASHISH-EATER;
OR, THE APOCALYPSE OF EVIL

Clark Ashton Smith

Bow down: I am the emperor of dreams;
I crown me with the million-colored sun
Of secret worlds incredible, and take
Their trailing skies for vestment when I soar,
Throned on the mounting zenith, and illume
The spaceward-flown horizons infinite.
Like rampant monsters roaring for their glut,
The fiery-crested oceans rise and rise,
By jealous moons maleficently urged
To follow me for ever; mountains horned
With peaks of sharpest adamant, and mawed
With sulphur-lit volcanoes lava-langued,
Usurp the skies with thunder, but in vain;
And continents of serpent-shapen trees,
With slimy trunks that lengthen league by league,
Pursue my flight through ages spurned to fire
By that supreme ascendance; sorcerers,
And evil kings, predominantly armed
With scrolls of fulvous dragon-skin whereon
Are worm-like runes of ever-twisting flame,
Would stay me; and the sirens of the stars,
With foam-like songs from silver fragrance wrought,
Would lure me to their crystal reefs; and moons
Where viper-eyed, senescent devils dwell,
With antic gnomes abominably wise,
Heave up their icy horns across my way.
But naught deters me from the goal ordained

By suns and eons and immortal wars,
And sung by moons and motes; the goal whose name
Is all the secret of forgotten glyphs
By sinful gods in torrid rubies writ
For ending of a brazen book; the goal
Whereat my soaring ecstasy may stand
In amplest heavens multiplied to hold
My hordes of thunder-vested avatars,
And Promethèan armies of my thought,
That brandish claspèd levins. There I call
My memories, intolerably clad
In light the peaks of paradise may wear,
And lead the Armageddon of my dreams
Whose instant shout of triumph is become
Immensity's own music: for their feet
Are founded on innumerable worlds,
Remote in alien epochs, and their arms
Upraised, are columns potent to exalt
With ease ineffable the countless thrones
Of all the gods that are or gods to be,
And bear the seats of Asmodai and Set
Above the seventh paradise.

                                        Supreme
In culminant omniscience manifold,
And served by senses multitudinous,
Far-posted on the shifting walls of time,
With eyes that roam the star-unwinnowed fields
Of utter night and chaos, I convoke
The Babel of their visions, and attend
At once their myriad witness. I behold
In Ombos, where the fallen Titans dwell,
With mountain-builded walls, and gulfs for moat,
The secret cleft that cunning dwarves have dug
Beneath an alp-like buttress; and I list,
Too late, the clang of adamantine gongs

Dinned by their drowsy guardians, whose feet
Have felt the wasp-like sting of little knives
Embrued with slobber of the basilisk
Or the pale juice of wounded upas. In
Some red Antarean garden-world, I see
The sacred flower with lips of purple flesh,
And silver-lashed, vermilion-lidded eyes
Of torpid azure; whom his furtive priests
At moonless eve in terror seek to slay
With bubbling grails of sacrificial blood
That hide a hueless poison. And I read
Upon the tongue of a forgotten sphinx,
The annulling word a spiteful demon wrote
In gall of slain chimeras; and I know
What pentacles the lunar wizards use,
That once allured the gulf-returning roc,
With ten great wings of furlèd storm, to pause
Midmost an alabaster mount; and there,
With boulder-weighted webs of dragons' gut
Uplift by cranes a captive giant built,
They wound the monstrous, moonquake-throbbing bird,
And plucked from off his saber-taloned feet
Uranian sapphires fast in frozen blood,
And amethysts from Mars. I lean to read
With slant-lipped mages, in an evil star,
The monstrous archives of a war that ran
Through wasted eons, and the prophecy
Of wars renewed, which shall commemorate
Some enmity of wivern-headed kings
Even to the brink of time. I know the blooms
Of bluish fungus, freaked with mercury,
That bloat within the craters of the moon,
And in one still, selenic hour have shrunk
To pools of slime and fetor; and I know
What clammy blossoms, blanched and cavern-grown,
Are proffered to their gods in Uranus

By mole-eyed peoples; and the livid seed
Of some black fruit a king in Saturn ate,
Which, cast upon his tinkling palace-floor,
Took root between the burnished flags, and now
Hath mounted and become a hellish tree,
Whose lithe and hairy branches, lined with mouths,
Net like a hundred ropes his lurching throne,
And strain at starting pillars. I behold
The slowly-thronging corals that usurp
Some harbor of a million-masted sea,
And sun them on the league-long wharves of gold—
Bulks of enormous crimson, kraken-limbed
And kraken-headed, lifting up as crowns
The octiremes of perished emperors,
And galleys fraught with royal gems, that sailed
From a sea-fled haven.

                      Swifter and stranger grow
The visions: now a mighty city looms,
Hewn from a hill of purest cinnabar
To domes and turrets like a sunrise thronged
With tier on tier of captive moons, half-drowned
In shifting erubescence. But whose hands
Were sculptors of its doors, and columns wrought
To semblance of prodigious blooms of old,
No eremite hath lingered there to say,
And no man comes to learn: for long ago
A prophet came, warning its timid king
Against the plague of lichens that had crept
Across subverted empires, and the sand
Of wastes that cyclopean mountains ward;
Which, slow and ineluctable, would come
To take his fiery bastions and his fanes,
And quench his domes with greenish tetter. Now
I see a host of naked giants, armed
With horns of behemoth and unicorn,

Who wander, blinded by the clinging spells
Of hostile wizardry, and stagger on
To forests where the very leaves have eyes,
And ebonies like wrathful dragons roar
To teaks a-chuckle in the loathly gloom;
Where coiled lianas lean, with serried fangs,
From writhing palms with swollen boles that moan;
Where leeches of a scarlet moss have sucked
The eyes of some dead monster, and have crawled
To bask upon his azure-spotted spine;
Where hydra-throated blossoms hiss and sing,
Or yawn with mouths that drip a sluggish dew
Whose touch is death and slow corrosion. Then
I watch a war of pygmies, met by night,
With pitter of their drums of parrot's hide,
On plains with no horizon, where a god
Might lose his way for centuries; and there,
In wreathèd light and fulgors all convolved,
A rout of green, enormous moons ascend,
With rays that like a shivering venom run
On inch-long swords of lizard-fang.

                                              Surveyed
From this my throne, as from a central sun,
The pageantries of worlds and cycles pass;
Forgotten splendors, dream by dream, unfold
Like tapestry, and vanish; violet suns,
Or suns of changeful iridescence, bring
Their rays about me like the colored lights
Imploring priests might lift to glorify
The face of some averted god; the songs
Of mystic poets in a purple world
Ascend to me in music that is made
From unconceivèd perfumes and the pulse
Of love ineffable; the lute-players
Whose lutes are strung with gold of the utmost moon,

Call forth delicious languors, never known
Save to their golden kings; the sorcerers
Of hooded stars inscrutable to God,
Surrender me their demon-wrested scrolls,
Inscribed with lore of monstrous alchemies
And awful transformations.

                             If I will,
I am at once the vision and the seer,
And mingle with my ever-streaming pomps,
And still abide their suzerain: I am
The neophyte who serves a nameless god,
Within whose fane the fanes of Hecatompylos
Were arks the Titan worshippers might bear,
Or flags to pave the threshold; or I am
The god himself, who calls the fleeing clouds
Into the nave where suns might congregate
And veils the darkling mountain of his face
With fold on solemn fold; for whom the priests
Amass their monthly hecatomb of gems—
Opals that are a camel-cumbering load,
And monstrous alabraundines, won from war
With realms of hostile serpents; which arise,
Combustible, in vapors many-hued
And myrrh-excelling perfumes. It is I,
The king, who holds with scepter-dropping hand
The helm of some great barge of orichalchum,
Sailing upon an amethystine sea
To isles of timeless summer: for the snows
Of hyperborean winter, and their winds,
Sleep in his jewel-built capital,
Nor any charm of flame-wrought wizardry,
Nor conjured suns may rout them; so he flees,
With captive kings to urge his serried oars,
Hopeful of dales where amaranthine dawn
Hath never left the faintly sighing lote

And lisping moly. Firm of heart, I fare
Impanoplied with azure diamond,
As hero of a quest Achernar lights,
To deserts filled with ever-wandering flames
That feed upon the sullen marl, and soar
To wrap the slopes of mountains, and to leap
With tongues intolerably lengthening
That lick the blenchèd heavens. But there lives
(Secure as in a garden walled from wind)
A lonely flower by a placid well,
Midmost the flaring tumult of the flames,
That roar as roars a storm-possessèd sea,
Implacable for ever; and within
That simple grail the blossom lifts, there lies
One drop of an incomparable dew
Which heals the parchèd weariness of kings,
And cures the wound of wisdom. I am page
To an emperor who reigns ten thousand years,
And through his labyrinthine palace-rooms,
Through courts and colonnades and balconies
Wherein immensity itself is mazed,
I seek the golden gorget he hath lost,
On which, in sapphires fine as orris-seed,
Are writ the names of his conniving stars
And friendly planets. Roaming thus, I hear
Like demon tears incessant, through dark ages,
The drip of sullen clepsydrae; and once
In every lustrum, hear the brazen clocks
Innumerably clang with such a sound
As brazen hammers make, by devils dinned
On tombs of all the dead; and nevermore
I find the gorget, but at length I find
A sealèd room whose nameless prisoner
Moans with a nameless torture, and would turn
To hell's red rack as to a lilied couch
From that whereon they stretched him; and I find,

Prostrate upon a lotus-painted floor,
The loveliest of all belovèd slaves
My emperor hath, and from her pulseless side
A serpent rises, whiter than the root
Of some venefic bloom in darkness grown,
And gazes up with green-lit eyes that seem
Like drops of cold, congealing poison.

                                              Hark!
What word was whispered in a tongue unknown,
In crypts of some impenetrable world?
Whose is the dark, dethroning secrecy
I cannot share, though I am king of suns,
And king therewith of strong eternity,
Whose gnomons with their swords of shadow guard
My gates, and slay the intruder? Silence loads
The wind of ether, and the worlds are still
To hear the word that flees mine audience.
In simultaneous ruin, all my dreams
Fall like a rack of fuming vapors raised
To semblance by a necromant, and leave
Spirit and sense unthinkably alone
Above a universe of shrouded stars
And suns that wander, cowled with sullen gloom,
Like witches to a Sabbath. . . . Fear is born
In crypts below the nadir, and hath crawled
Reaching the floor of space, and waits for wings
To lift it upward like a hellish worm
Fain for the flesh of cherubim. Red orbs
And eyes that gleam remotely as the stars,
But are not eyes of suns or galaxies,
Gather and throng to the base of darkness; a flame
Behind some black, abysmal curtain burns,
Implacable, and fanned to whitest wrath
By raisèd wings that flail the whiffled gloom,
And make a brief and broken wind that moans

As one who rides a throbbing rack. There is
A Thing that crouches, worlds and years remote,
Whose horns a demon sharpens, rasping forth
A note to shatter the donjon-keeps of time,
Or crack the sphere of crystal. All is dark
For ages, and my tolling heart suspends
Its clamor as within the clutch of death
Tightening with tense, hermetic rigors. Then,
In one enormous, million-flashing flame,
The stars unveil, the suns remove their cowls,
And beam to their responding planets; time
Is mine once more, and armies of its dreams
Rally to that insuperable throne
Firmed on the zenith.

                    Once again I seek
The meads of shining moly I had found
In some anterior vision, by a stream
No cloud hath ever tarnished; where the sun,
A gold Narcissus, loiters evermore
Above his golden image. But I find
A corpse the ebbing water will not keep,
With eyes like sapphires that have lain in hell
And felt the hissing coals; and all the flowers
About me turn to hooded serpents, swayed
By flutes of devils in lascivious dance
Meet for the nod of Satan, when he reigns
Above the raging Sabbath, and is wooed
By sarabands of witches. But I turn
To mountains guarding with their horns of snow
The source of that befoulèd rill, and seek
A pinnacle where none but eagles climb,
And they with failing pennons. But in vain
I flee, for on that pylon of the sky
Some curse hath turned the unprinted snow to flame—
Red fires that curl and cluster to my tread,

Trying the summit's narrow cirque. And now
I see a silver python far beneath—
Vast as a river that a fiend hath witched
And forced to flow reverted in its course
To fountains whence it issued. Rapidly
It winds from slope to crumbling slope, and fills
Ravines and chasmal gorges, till the crags
Totter with coil on coil incumbent. Soon
It hath entwined the pinnacle I keep,
And gapes with a fanged, unfathomable maw
Wherein great Typhon and Enceladus
Were orts of daily glut. But I am gone,
For at my call a hippogriff hath come,
And firm between his thunder-beating wings
I mount the sheer cerulean walls of noon
And see the earth, a spurnèd pebble, fall—
Lost in the fields of nether stars—and seek
A planet where the outwearied wings of time
Might pause and furl for respite, or the plumes
Of death be stayed, and loiter in reprieve
Above some deathless lily: for therein
Beauty hath found an avatar of flowers—
Blossoms that clothe it as a colored flame
From peak to peak, from pole to sullen pole,
And turn the skies to perfume. There I find
A lonely castle, calm, and unbeset
Save by the purple spears of amaranth,
And leafing iris tender-sworded. Walls
Of flushèd marble, wonderful with rose,
And domes like golden bubbles, and minarets
That take the clouds as coronal—these are mine,
For voiceless looms the peaceful barbican,
And the heavy-teethed portcullis hangs aloft
To grin a welcome. So I leave awhile
My hippogriff to crop the magic meads,
And pass into a court the lilies hold,

And tread them to a fragrance that pursues
To win the portico, whose columns, carved
Of lazuli and amber, mock the palms
Of bright Aidennic forests—capitalled
With fronds of stone fretted to airy lace,
Enfolding drupes that seem as tawny clusters
Of breasts of unknown houris; and convolved
With vines of shut and shadowy-leavèd flowers
Like the dropt lids of women that endure
Some loin-dissolving ecstasy. Through doors
Enlaid with lilies twined luxuriously,
I enter, dazed and blinded with the sun,
And hear, in gloom that changing colors cloud,
A chuckle sharp as crepitating ice
Upheaved and cloven by shoulders of the damned
Who strive in Antenora. When my eyes
Undazzle, and the cloud of color fades,
I find me in a monster-guarded room,
Where marble apes with wings of griffins crowd
On walls an evil sculptor wrought; and beasts
Wherein the sloth and vampire-bat unite,
Pendulous by their toes of tarnished bronze,
Usurp the shadowy interval of lamps
That hang from ebon arches. Like a ripple
Borne by the wind from pool to sluggish pool
In fields where wide Cocytus flows his bound,
A crackling smile around that circle runs,
And all the stone-wrought gibbons stare at me
With eyes that turn to glowing coals. A fear
That found no name in Babel, flings me on,
Breathless and faint with horror, to a hall
Within whose weary, self-reverting round,
The languid curtains, heavier than palls,
Unnumerably depict a weary king
Who fain would cool his jewel-crusted hands
In lakes of emerald evening, or the fields

Of dreamless poppies pure with rain. I flee
Onward, and all the shadowy curtains shake
With tremors of a silken-sighing mirth,
And whispers of the innumerable king,
Breathing a tale of ancient pestilence
Whose very words are vile contagion. Then
I reach a room where caryatides,
Carved in the form of voluptuous Titan women,
Surround a throne flowering ebony
Where creeps a vine of crystal. On the throne
There lolls a wan, enormous Worm, whose bulk,
Tumid with all the rottenness of kings,
Overflows its arms with fold on creasèd fold
Obscenely bloating. Open-mouthed he leans,
And from his fulvous throat a score of tongues,
Depending like to wreaths of torpid vipers,
Drivel with phosphorescent slime, that runs
Down all his length of soft and monstrous folds,
And creeping among the flowers of ebony,
Lends them the life of tiny serpents. Now,
Ere the Horror ope those red and lashless slits
Of eyes that draw the gnat and midge, I turn
And follow down a dusty hall, whose gloom,
Lined by the statues with their mighty limbs,
Ends in a golden-roofèd balcony
Sphering the flowered horizon.

                                    Ere my heart
Hath hushed the panic tumult of its pulses,
I listen, from beyond the horizon's rim,
A mutter faint as when the far simoon,
Mounting from unknown deserts, opens forth,
Wide as the waste, those wings of torrid night
That shake the doom of cities from their folds,
And musters in its van a thousand winds
That, with disrooted palms for besoms, rise,

And sweep the sands to fury. As the storm,
Approaching, mounts and loudens to the ears
Of them that toil in fields of sesame,
So grows the mutter, and a shadow creeps
Above the gold horizon like a dawn
Of darkness climbing zenith-ward. They come,
The Sabaoth of retribution, drawn
From all dread spheres that knew my trespassing,
And led by vengeful fiends and dire alastors
That owned my sway aforetime! Cockatrice,
Python, tragelaphus, leviathan,
Chimera, martichoras, behemoth,
Geryon, and sphinx, and hydra, on my ken
Arise as might some Afrit-builded city
Consummate in the lifting of a lash
With thunderous domes and sounding obelisks
And towers of night and fire alternate! Wings
Of white-hot stone along the hissing wind
Bear up the huge and furnace-hearted beasts
Of hells beyond Rutilicus; and things
Whose lightless length would mete the gyre of moons—
Born from the caverns of a dying sun—
Uncoil to the very zenith, half-disclosed
From gulfs below the horizon; octopi
Like blazing moons with countless arms of fire,
Climb from the seas of ever-surging flame
That roll and roar through planets unconsumed,
Beating on coasts of unknown metals; beasts
That range the mighty worlds of Alioth rise,
Afforesting the heavens with multitudinous horns
Amid whose maze the winds are lost; and borne
On cliff-like brows of plunging scolopendras,
The shell-wrought towers of ocean-witches loom;
And griffin-mounted gods, and demons throned
On sable dragons, and the cockodrills
That bear the spleenful pygmies on their backs;

And blue-faced wizards from the worlds of Saiph,
On whom Titanic scorpions fawn; and armies
That move with fronts reverted from the foe,
And strike athwart their shoulders at the shapes
Their shields reflect in crystal; and eidola
Fashioned within unfathomable caves
By hands of eyeless peoples; and the blind
Worm-shapen monsters of a sunless world,
With krakens from the ultimate abyss,
And Demogorgons of the outer dark,
Arising, shout with dire multisonous clamors,
And threatening me with dooms ineffable
In words whereat the heavens leap to flame,
Advance upon the enchanted palace. Falling
For league on league before, their shadows blight
And eat like fire the amaranthine meads,
Leaving an ashen desert. In the palace
I hear the apes of marble shriek and howl,
And all the women-shapen columns moan,
Babbling with terror. In my tenfold fear,
A monstrous dread unnamed in any hell,
I rise, and flee with the fleeing wind for wings,
And in a trice the wizard palace reels,
And spiring to a single tower of flame,
Goes out, and leaves nor shard nor ember! Flown
Beyond the world upon that fleeing wind
I reach the gulf's irrespirable verge,
Where fails the strongest storm for breath, and fall,
Supportless, through the nadir-plungèd gloom,
Beyond the scope and vision of the sun,
To other skies and systems.

                                   In a world
Deep-wooded with the multi-colored fungi
That soar to semblance of fantastic palms,
I fall as falls the meteor-stone, and break

A score of trunks to atom-powder. Unharmed
I rise, and through the illimitable woods,
Among the trees of flimsy opal, roam,
And see their tops that clamber hour by hour
To touch the suns of iris. Things unseen,
Whose charnel breath informs the tideless air
With spreading pools of fetor, follow me,
Elusive past the ever-changing palms;
And pittering moths with wide and ashen wings
Flit on before, and insects ember-hued,
Descending, hurtle through the gorgeous gloom
And quench themselves in crumbling thickets. Heard
Far off, the gong-like roar of beasts unknown
Resounds at measured intervals of time,
Shaking the riper trees to dust, that falls
In clouds of acrid perfume, stifling me
Beneath an irised pall.

                    Now the palmettoes
Grow far apart, and lessen momently
To shrubs a dwarf might topple. Over them
I see an empty desert, all ablaze
With amethysts and rubies, and the dust
Of garnets or carnelians. On I roam,
Treading the gorgeous grit, that dazzles me
With leaping waves of endless rutilance,
Whereby the air is turned to a crimson gloom
Through which I wander blind as any Kobold;
Till underfoot the grinding sands give place
To stone or metal, with a massive ring
More welcome to mine ears than golden bells
Or tinkle of silver fountains. When the gloom
Of crimson lifts, I stand upon the edge
Of a broad black plain of adamant that reaches,
Level as windless water, to the verge
Of all the world; and through the sable plain

A hundred streams of shattered marble run,
And streams of broken steel, and streams of bronze,
Like to the ruin of all the wars of time,
To plunge with clangor of timeless cataracts
Adown the gulfs eternal.

                So I follow
Between a river of steel and a river of bronze,
With ripples loud and tuneless as the clash
Of a million lutes; and come to the precipice
From which they fall, and make the mighty sound
Of a million swords that meet a million shields,
Or din of spears and armor in the wars
Of half the worlds and eons. Far beneath
They fall, through gulfs and cycles of the void,
And vanish like a stream of broken stars
Into the nether darkness; nor the gods
Of any sun, nor demons of the gulf,
Will dare to know what everlasting sea
Is fed thereby, and mounts forevermore
In one unebbing tide.

                      What nimbus-cloud
Or night of sudden and supreme eclipse,
Is on the suns of opal? At my side
The rivers run with a wan and ghostly gleam
Through darkness falling as the night that falls
From spheres extinguished. Turning, I behold
Betwixt the sable desert and the suns,
The poisèd wings of all the dragon-rout,
Far-flown in black occlusion thousand-fold
Through stars, and deeps, and devastated worlds,
Upon my trail of terror! Griffins, rocs,
And sluggish, dark chimeras, heavy-winged
After the ravin of dispeopled lands,
And harpies, and the vulture-birds of hell,

Hot from abominable feasts, and fain
To cool their beaks and talons in my blood—
All, all have gathered, and the wingless rear,
With rank on rank of foul, colossal Worms,
Makes horrent now the horizon. From the van
I hear the shriek of wyvers, loud and shrill
As tempests in a broken fane, and roar
Of sphinxes, like relentless toll of bells
From towers infernal. Cloud on hellish cloud
They arch the zenith, and a dreadful wind
Falls from them like the wind before the storm,
And in the wind my riven garment streams
And flutters in the face of all the void,
Even as flows a flaffing spirit, lost
On the pit's undying tempest. Louder grows
The thunder of the streams of stone and bronze—
Redoubled with the roar of torrent wings
Inseparably mingled. Scarce I keep
My footing in the gulfward winds of fear,
And mighty thunders beating to the void
In sea-like waves incessant; and would flee
With them, and prove the nadir-founded night
Where fall the streams of ruin. But when I reach
The verge, and seek through sun-defeating gloom
To measure with my gaze the dread descent,
I see a tiny star within the depths—
A light that stays me while the wings of doom
Convene their thickening thousands: for the star
Increases, taking to its hueless orb,
With all the speed of horror-changèd dreams,
The light as of a million million moons;
And floating up through gulfs and glooms eclipsed
It grows and grows, a huge white eyeless Face
That fills the void and fills the universe,
And bloats against the limits of the world
With lips of flame that open. . . .

Clark Ashton Smith

Visions of Golconda

Richard L. Tierney

Now comes the night with visions wild and fair,
With grandiose vistas of unending wealth.
My mind dissolves in drug-begotten dreams;
Swift from my blackened brain spill out the thoughts
Of stifling poverty and impotence,
Like swarthy bats that pour in legions vast
From nighted caverns 'neath Southwestern hills,
And in their place come scenes of fiery riches—
Fantastic glimmerings of shifting gold
And hills of rolling gem-stones million-hued,
Advancing in unending ranks upon
A plain whose far horizon stretches past
The limits of the eye. What dreams are these
Where wealth beyond the hopes of all mankind—
Beyond the dreams of all earth's ages past
And ages yet to come—is mine!

                      Behold!
I stand upon a sea of rippling opal
Bright-polished, dazzling to my gleeful gaze,
Reflecting rays from multitudes of moons
Whose multi-colored discs illumine bright
The billion facets of its broken plain.
Entranced, I wander through this world of wealth,
Alone, engrossed in visionings of power,
Plucking from out innumerable cracks
And crevices where they have haply lodged
Huge diamonds as clear as mountain lakes
And cut to a perfection far above

The handiwork achieved by tools of men.
I revel in an ocean of such wealth—
Such panoramas of all treasures known—
That all my thoughts dissolve in gleeful chaos,
Leaving me free to glory in my power
Unhindered by the minds of other men,
Unchecked by all their vain moralities.
Now on I stride, as with the wingéd shoes
Of Mercury, swift-splashing over dunes
Of silver coins up-piled to form a beach
Beside the sea of opal—my domain
Whose endless reaches hold no elements
Save those of untold value.

      On I march,
Surveying all the gleaming, silver span
Of that great gleaming beach, where shifting waves
Rise up as toppling piles of sparkling gems—
Rubies and emeralds—aquamarines—
Clashing like crystal marbles by the tons,
Splashing in tinkling waves upon the coins,
Shifting, retreating, rolling down the sand
Of silver to the sea from whence they came,
Leaving upon the glittering expanse
Like bright wave-stranded shells upon a beach,
Great gleaming ornaments of flaming gold
In shapes fantastical and demon-wrought.
Enthralled, I comb this strand with avid joy,
Striving in vain to heft the ponderous drift—
Monstrous medallions, jewel-studded gods
Molded of old in blazing crucibles
By sorcerers of Oriental lands,
Exotic figurines by Eros cast
As offerings into these seas of wealth,
Whose naiad-shapen forms shed such a spell
O'er any with the fortune to behold
As to awaken dreams of ecstasy

And longings that their bodies of hard gold
Might soften with the throbbing pulse of life
And change to warming flesh.

                                    But now I turn
My back upon this roaring sea of gems
To gaze across the towering continent
Whose metal shore holds back the flashing waves.
With wonder I behold the needle-spires
Of mountains whose high, heaven-piercing peaks
Jut out beyond the scented atmosphere
That girds this mighty planet where the wealth
From untold worlds has gathered on the tides
Of gravity to form a sparkling sphere
Whose weight of wealth would crush man's world to dust.
My eyes grow dazzled by the flaring gleam
That leaps from out those scintillating spires
Wrought of pure emerald by forces vast
In some forgotten age of seismic strife.
Their fanglike peaks reflect the feeble rays
Of wheeling moons whose orbs of platinum
Scrape gently on the mountains' needle-horns
As they pursue their orbits. In their beams
I see a desert stretching off between
The opal sea and crystal mountains vast,
Whose sand is formed by discs of ancient gold
Wherein stand rooted, gaunt and cacti-like,
Titanic corals of such splendid worth
That one could purchase all the realms of kings,
Or all the titles of nobility
That man grants to his vanity and pride,
Or passage to those paradisial realms
Beyond the gates of purgatory.

                                            Hark!
What is this dinful roar that smites my ears
From out beyond the many-colored sea?

Richard L. Tierney

What is this light that bursts upon my eyes,
Blinding my vision with supernal glare,
Drowning the world in an effulgent haze?
Now, lo! my tearful orbs, adjusting, spy
A rising sun of sapphire looming vast
Above the opal ocean's wide expanse.
How bright its rays of liquid violet
Pour from its flaming heart upon the sea,
Turning the sullen opal's flickerings
To fiery flashings wild and serpentine
Like mirrored, half-seen essences that lurk
Within some witch-brewed potion cauldron-stirred!
Now even as I watch I see it climb
Into the flaring heavens, while the din
Grows louder from the gem-encrusted sea,
Advancing like a monstrous tide-drawn wave
With thunders like to that of Doomsday's peal.
And still, with billion facets blazing fierce
As bursting atoms swelling with blue power,
The sapphirean sun climbs up the sky
And arcs with ponderous curve its mounting course.
Now zenithward it soars; I hear the crash
Of toppling cliffs of gemstones; now I see,
Far out upon the opal-bright expanse
Where dim horizons merge with violet sky,
A cliff-high wave of diamonds looming white—
A crest of sparkling crystal foam that grows
And rumbles in a menacing advance,
Hurled on its course by unseen, towing bonds
From that blue sun in whose high wake it flows.
Nearer it looms with soul-inspiring roar,
Devouring all horizons with its bulk,
Advancing with a speed that bodes of doom
While, hypnotized, I watch its dread approach—
Its towering ramparts surge against the shore—
Its mile-high crest rain down like shattered ice
To swallow up the universe . . .

Memoria:
A Fragment from the Book of Wyvern

Leigh Blackmore

I

Strong anodyne and opiate I quaff
To lull me slowly from the waking world;
The air hangs heavy, hot and drugged, as through
The mires of consciousness my sluggish mind
Sinks, ennui-swamped, to finally repose
Within the easeful lands beyond the Earth
Where often I have sought to calm my soul.
Lethargic movements cease and drowsy brain
Revives as glittering motes of dream-dust dance
Before me, coalesce to mirror thought,
And part, revealing lustrous marble walls
I knew before in lives long past. Behold
The city where aeons ago I dwelled,
And trained under the wizard's dark-eyed gaze
To reach the perfect peak of sorcery
In that forbidden quarter where alone
The seers and mages built their basalt towers
And cast their potent horoscopes and spells.
I enter through the giant brazen gates
And stand within the vast courtyard alone,
Where once the chimes of sapphire-spired fanes
Were sweetly heard and merchants from far lands
Displayed exotic wares to vanished crowds.
Around me, ornate balconies are joined
By stairs that spiral, ornamental walks

Whose balustrades and colonnades lead on
The eye to arches delicate and fine,
Resplendent domes and columned porticoes.
Tall, slender towers topped by minarets
Of coral rise from well-proportioned streets
Where polished statues of chalcedony
Stand gracefully amidst islets of green
Trim lawns, and fragrant gardens, zephyr-cooled,
Lie sleepy 'neath the sun's vermilion rays;
Proud peacocks spread their tails with shining eyes
Like purple jewels, strut regally past
Young trees in lovely orchards by the road.
Ripe, softly glowing fruits profusely hang
Among the drooping branches; on the air
Float scents of blossom-laden cherry boughs.
I move past silver fountains spraying still
Their light polychromatic water-jets
And leave behind the sector's opulence
To travel swiftly by well-known wide streets
Unto the onyx-paved sorcerer's realm.

II

In youth, I first espied the darkling manse
Where dwelled th' enchanter, he who reigned supreme,
And braved the dangers of its bounds to gain,
Through him, apprenticeship to evil gods;
Have dominion o'er all things natural
And occult by black art, and e'en command
That Source from which no man at last returns.
Beneath the carmine skies of this dread place,
I learned to raise the spirits of the dead
And take their secret knowledge unto me;
From rune-writ tomes of eldritch, antique lore
My mentor taught me all those formulae
By which man's brain, else feeble and infirm,

Can be borne up supernal 'gainst the plans
Of daemons baleful and maleficent
Which seek to overthrow magicians' power.
I conjured fiery angels and the rank
And file of Hell's hierarchy terrible;
Evoked (by rituals complex and reviled)
Great wheeling golden Phoenixes from out
The precincts of the sun; upon strange sands
I inscribed planet signs and summoned up
The lich who found the polished mirror bronze
Of sunken Poseidonis; I burned herbs
And incense from pure magic oils distilled
That necromantic servitors brought back
From Aegypt's shadowed tombs and temples dead;
And after years, did build an aedifice
Of porphyry and edomite within
The sorcerers' domain to take my own
Share of the treasures of the world without.

III

Now am I there once more at fall of night,
And as I gaze upon remembered things—
The blasting-wand and brazier, deviced
With lions winged and serpents venom-fanged—
Music by Winter's frosty wind-glance blown
Beats airy fists on mirror panes nearby
And I peer out to see the sunless globes
Of thick, dark night give way to clear, cold skies.
Light-rings, like blazing dragon-scalèd eyes,
Float there among the planets, each to each
Mist-dripping golden rain like running fire,
Deep-drowning hills whose snowy peaks uprise
From valleys rubbed like copper candlesticks
With moonshine; and ice-crystal clouds, illumed
By starry lanterns, snow the silent plain

With drifting flakes. Afar I hear a sound
Of footsteps manifold which nearer grows,
A muffled tread of iron gods below
And bronze monarchs striding past in the snow,
Making a file of sea-swept helmet plumes.
Their crusted garments flow beneath full beards
And faces leprous-white with dire portent;
They clasp great volumes, antient, vellum-bound,
And pentacles and rose-quartz chalices.
It is th' assembled might of all the town,
Combined for some reason arcane, who march,
Phantoms whose ivory staffs leave silver trails
On tilèd streets and gleaming ice-glazed roads
Unto the water's edge, and start to chant.
The rising song, the orison of power
And fearful malediction imminent,
Now strikes a chord in distant memory—
Was known to me, but lies beyond my reach.
The scarlet-streakèd sky pulsates with sound
Subliminal; a dull, low thudding beats
Upon my ears, assails my senses all.
Throbs, as of some unending heartbeat, move
The heavens (filled with amaranthine light)
To violent turmoil. Veils of crimson cloud
Tear wild across the blazing welkin; fires
Outpour the rift in opal-flashing streams
Of dazzling brilliance which transform the lands
Beneath, throw em'rald bolts on hill and vale,
Let flow prismatic floods on beryl seas,
Transmute the air to stained glass rare, and make
Aethereal gold o'erlie the moon's light side.
The iridescent landscape trembles, shakes;
The seas grow huge and hollow; massive waves
Of coruscating colours roar and pound
Upon the barren shore's wide sweep. The boom
Of surf and sky in unison with that

Arising from commanding sorcerers
Builds in intensity, cuts through the night
To hammer at the tower in which I stand.
The floor beneath me rocks; the giant panes
Begin to crack; the walls—rent stone from stone—
Gape wide and tumble; chill wind rushes in,
As does recall. I scream as knowledge comes
And I remember where I heard that chant—
The time I died before, as now I do,
Intruder in this realm. The world explodes.
Chill star-winds howl around me as I fall
Fear-frozen in the thunderous abyss
Whose gape rends wide the vacuum's fulgent stuff,
The spatial plane, and full reveals the glint
Of galaxies beyond our solar orb;
Star-shattered panorama of dead worlds
Where dust and rock alone behold the course
Of other planets, stricken, reeling down
The corridors of empty space remote,
To grind, and slow, and lifeless float; or flare
In brief vitality of death-throes huge
And inconceivable in scale and then
To drift away as primal cosmic dust . . .

A Trip to the Hypnotist

Alan Gullette

The sign read "Dr Spiegel, Hypnotist"
And so I went into the waiting room—
Devoid of patients, not of magazines—
And sat myself upon an empty chair.

They said you could be hypnotized and see
Before your eyes the lives that you had lived
Before this life. Although I had my doubts,
I went to know if I had lived before.

When it was time, I went into a room
Half-dark, half-lit by glowing amber light.
"Just watch my pocket watch," he said to me,
And swung the gold watch on its silver chain.

The watch swung like a pendulum, and then
Became a shining, golden orb—a sun
That moved across a blackness specked with stars
That flickered, faded, fell—and dark was all . . .

"Remember when you were a little boy—
A time of peace, of lazy afternoons
Before the toils of life wrapped you in coils . . ."
And I remembered.

                    When I was a child,
We lived alone—a mansion in the woods,
A place remote—my great-grandfather's house.
He was an ancient man, both wise and strange
And lived among his books of ancient lore.

Some said he was a wizard, some a fool.
One day there was a visitor . . . a man
That looked like Dr Spiegel, only young.
I tried to speak, but could not find my voice.

"Now go back further, when you were a babe.
Your mother nursed you, bathed you every day.
Go further still—when you were in the womb,
Afloat in darkness . . ."

                                              And I floated there,
Borne on the inner tide until my birth,
A timeless span in sanguine twilight spent
Aware of nothing but the beating heart,
Dim songs, sharp laughter, and the wail of tears.

"Now, go back further, back before your time,
Before your father and your mother lived.
Do you recall? You were *another* then,
A man of boundless power—and a sage."

Through haze of years I gazed . . . My vision cleared.
I looked into a looking glass and saw
The face of my great-grandfather, not me!
I wore his clothes, about me were his things . . .

"Relax—fear not, my son—remain with me,"
A voice that was not mine spoke from my lips.
"I have not died, but wait to live again.
Join me and we will live *for centuries* . . ."

I fled in mortal fear—fled back through time,
Through lives of ancestors of ancestors,
Lives spent in England and in France, in Spain,
In Rome, in ancient Greece, in Tyre, in Ur . . .

I was an orator, a serf, a scribe,
A chief, a warrior and an artisan.

I saw great cities rise and saw them fall
In different lives—as woman, man, and child.

Each time, I was pursued by nameless fear—
A voice that called my name within my head,
An evil thing that sought to rule my mind—
Back through the lives recorded in my genes . . .

Back to a time when I was not a man,
But some small ape that crept beneath the trees
Hid out in caves and feared the phasing moon,
That held its head and shrieked into the night!

And hounded still, I passed through countless lives:
A shrew that scampered, dodging taloned wings;
An asp that slithered on and in the earth;
A fish that foundered on the surging sea . . .

And on I went, to microscopic realms
To squirm and wriggle in a tidal pool,
Till I became again a single cell—
With which, they say, all life we know began . . .

But brewing in that sea, an evil thought
Awoke in dim beginnings of the mind
That I was not myself—but someone else—
And sent me reeling further back in time

When molecules first formed, their atoms bound
Together by electrons shared through space:
Dual orbits, like a moon that moves between
Two planets, side by side . . .

                                        But then, repulsed
By some dark drive or fear, I broke away,
Went deeper still, resigned to break apart
In subatomic particles at last:
The spheres-in-spheres that spin and form this world.

What fields of being crowd and intersect
Within the tilting planes that crash, collide
Like galaxies! I saw the Milky Way
Engorged with stars; I saw our star, the Sun . . .

And so I made my way to Earth again,
Returned, and watched the world spin round and round.
It swung on a chain wound around the sun—
A silver chain that links and binds us all . . .

When I awoke, the first thing that I saw
Was Dr Spiegel's dark, horrific face—
Protruding tongue, white lips and bulging eyes.
Around his neck the silver chain was wrapped;

The pocket watch now dangled at his breast.
I fled the room—a side door to the hall—
And down the empty corridor I dashed . . .
And I was free to go—go on my way.

Thirteen Ways of Looking At and Through Hashish

Bruce Boston

(After Clark Ashton Smith and Wallace Stevens)

The Art

Women Smoking Hashish, 1887,
by Italian Pointillist/Symbolist
Gaetano Previati, 1852 – 1920,
is a drab and moralistic
interior landscape,
executed in shades
of dull brown
grained and splashed
by pale yellow,
a bit like the shade
of fine Moroccan hash.

Four women slouch
on chairs and a couch,
dressed heavily
in the long layered
style of the era,
heads thrown back
in stuporous enchantment,
eyes closed
beneath a low-ceilinged room.

Here is a vision
opaque as the canvas
on which it is painted.

At the rear of the room,
through windows stained
by indelible smoke,
the strained light
fails to illuminate
the scene beyond.

The women look
as if they could
never stir again
and remain satisfied
unto death.

In all of this
depressing tableau,
there is no hint
of the colors
that obsess their minds,
nor the visions
that now consume
their lives.

The Rush

you know   man
this is just the way
it comes on for me

probably different
for everyone

when I smoke hash
you know   that first rush
it comes on like
no other drug

you ever been to
one of those classical concerts

with the big orchestras

well   hash
is like that orchestra
from one of those concerts
tuning up in the orchestra pit

that's how it comes on
when I smoke it
I like to call it
the tuning fork of hash

man   it's strange
and kind of awesome
the way it comes on at first

all the instruments making
different noises that
have nothing to do with
one another

like you can feel
the different parts
of your brain
tuning up
for the high like that

flaring up a little at a time
and getting
kind of all synchronized
and ready take off
together

and then gradually
one by one
they die down
to a steady glow

and there is a moment
or two of silence

no   almost silence

and you know —
to continue this riff
— maybe there is
a cough from the audience
the rustle of people
shifting in their chairs —
or maybe it's like
a rifling of pages
through your mind
too fast to read

and then it starts
hits you all at once
like a brush fire

takes the lid off
the top of your head
like some Crumb cartoon

and you are eight miles high
and all kinds of things
coming rushing in

and you can become
just about anything
and be anywhere
you want to be

The Flight

I am the rider
of the silken beast

of passionate hallucination.
I am the beast itself.

I grasp my mane
over twenty snowy mountains
and twenty flaming rivers.

I am the emperor
of red imagination
and ice cream dreams.

I am the black bird
whose shadow wing
sweeps the moon
in its flickering embrace,

the midnight vision
that slyly haunts
your shallow afternoons,

a wild storm of shades
that returns with evening,
nourished and aroused.

I am the fierce raptor
whose swan song
you have yet to hear.

The Victim

Assaulted by chords
Of apocalyptic thunder
Gliding forth from
The deepest pits
Of some Stygian hell,
Chords of fey emancipation
And wild enslavements,
Enabled by demons

Conversing in a humid
Garbled gutter tongue
As they debate my fate,
I am of a single mind
Of manifold illusions,
A tree whose branches
Never stop sprouting
Leaves and flowers
In networks interlacing,
Immersed in a morphing world
Of glyphs and ruins,
And amaranthine terrains,
Adrift in rich persecutions
And megalomanias,
I am crowned The King
Of My Own Distorted Dreams.

Passage of the Beast

I rush down dark tunnels with filthy water slogging
about my boots, the jagged cave walls about me

embedded with a luminous aromatic fungi casting
a pale odiferous light that sparks diamond flecks

and wafts the aroma of burnt cinnamon and sugar,
an ominous ocher light that ripples and undulates

with amorphous and anthropomorphic shapes.
Baroque encrustations accelerate toward me,

lost impresa embossed with intaglios like brands.
The archaeological debris of a thousand empires

coagulates in the corridors of my mind and I
embrace the hollow rites of ritual slaughter,

make pacts with demons of my shattered soul.

I fashion barbarian homunculi warriors who

writhe forth from the dark soil at my bidding.
I plunder riches from the hoards of my enemies.

I ride the helpless virgins like Khan or Attila,
seizing mates from the conquered hordes.

Once I emerge on the surface of the earth
the world is transformed to my bidding.

A glass carriage iridescent with northern lights
appears before me, its lucent seats cushioned

with extravagant pillows of liquid burgundy.
I embark on a journey that seems without end,

sailing over seascapes, landscapes, cityscapes.
Now it is dawn across this world of my making.

I hear black birds cawing the chorus of the hunt.
I hear their strident song and I see their shadows

swift against the sky and on the slanted streets.
I listen as they sing the dark legend of my life.

The Hashshashin

In the nanosecond
left to him
and his life,
as the bomb detonated,
he realized there
would be no paradise,
no virgins, no smoke.

The Chemistry

let me tell you
about this hash
that you are going
to want to buy
from me

ever heard of trichomes?
that's where the action is

they're these glandular hairs
that stick up from the buds
and flowers of the female
cannabis sativa plant

that's where the THC
the tetrahydrocannabinol
is most concentrated
and will take you
wherever you
want to go

cause you got these
cannabinoid receptors
located on neurons
scattered throughout
your brain and body
(just sitting there
ready and waiting
to get turned on)

and once the THC
binds with those receptors
your synapses start clicking
like a chain reaction
and that's when you
really begin to fly

now most hash
it's got maybe
thirty forty percent
trichomes mixed
in with a bunch
of leaves and flowers

but this hash
that you're going
to want to buy
from me
is Moroccan hash
the very best
and it's got sixty
percent guaranteed

you take one or two
hits and it's all
you're going to
ever need to fly
wherever you
want to go

the only question is
how much do you
want to buy?

The Gourmand

She ate furry trichomes
and fly agaric
from the cups
of flamboyant mushrooms.

She devoured rubies
of the finest persuasion
and emeralds
from the hearts of lizards.

She swallowed
diamonds drenched
in the aqua vitae
of extraterrestrial streams,

consumed spatulate
leaves of unknown origin
and incredible effect.
She graced the

tasteless crackers of the Host
with truffles and brie,
with fungal infestations
bred in the cellars

of decadent aristocrats,
washed them down with
nectar of absinthe
and sun stroked afternoons.

She drowned diligently
in the inimitable
pleasures of bizarre taste
and visionary extravaganza.

The Synesthesia

Shadow lines of night
close on the horizon,
slicing the last sounds of light
to strips so thin
they are ultrasonic,

like finely hammered
gold leaf
pressed
to the thickness of a single atom,
so thin they are transparent

to the inner ear
in the yellow lamps of dusk,

so thin you
can almost hear
their translucent shades,
taste their fragrance
on the tines of tomorrow.

The Flight of Language

I experience enchantments
of mythic proportions.
I am the owl and the raven,

the kingfisher, the heron,
the eagle and the hawk.
All birds of prey

forever in flight
or about to take flight,
all black birds in silhouette

against a harsh horizon,
diverse hybrids
of the same inky strain.

I explode to fractal feathers
beneath a semiotic sky
engraved with cloud runes

and clouds glyphs
in a language arcane
and illuminating.

As if words were riven
by endless dichotomies,
an ongoing dialectic,

each thought entrenched
and bastioned by others,
beleaguered by innuendo

and extended hyperbole,
lodged as a riddle
in a complex puzzle box,

the aged grain of its wood
darkened and polished
through the centuries

by hands that have
tried to unravel
its wiredrawn intricacy,

by minds that have
tried to unhinge
the sky.

The History

let me tell you
when you eat hash
when you smoke hash
you become part
of a grand tradition
of enhanced consciousness
and individual expression
stretching back centuries

thousands of years ago
let's say   ten or so
this cat somewhere
in the Himalayas
gets real hungry

no food anywhere

so he decides to eat some
leaves and flowers
off this weed growing
just about everywhere

now   they don't do
much for his hunger
but they do a lot
of things for his head
so he shares them
with his friends

now   time trip with me
for a minute or two
several thousand years
to be exact

the word has spread
in every direction
and so has the trade
all round Asia
east to China
south to India

it's not called a weed
anymore but sacred grass,
bhang, bhangi, keef, charas

Shiva   that Indian avatar
is dubbed the Lord of Bhang

meanwhile   this other cat
(and who knows
maybe he's a descendent
of that first cat?)
starts fooling around
until he finds the best

parts of sacred grass
and from these
he conjures
the first hashish

and pretty soon all
those ancient cats
are smoking it
eating it   smoking it

Scythians, Persians
Siberians, Samaritans,
Hiptherians, Dylusians,
you name
some ancient folk
from thereabouts
and you can bet
some of them
were getting high

now   stay with me here
jumping a few more years
and getting specific

about the time of Jesus
the first stash boxes
start showing up
coincidence?

900 – 1000
hashish spreads swiftly
through all of Arabia
and creeps into Europe
on a shady afternoon

1378
the Emir of the Ottoman Empire
issues a ban against hashish
throughout his kingdom
yeah   even back then
they were shooting
down the dreamers

mid-16th century
Arabian poet Mohammad Foruli
writes a long allegorical poem
about a battle between
hashish and wine
neither wins

1798
after conquering Egypt
Napoleon bans
the use of hash there
though his troops
carry it home

45 years later
le Club des hachichins
is founded
by the literati of France
(Dumas, Baudelaire,
Gautier, Nerval)
and opens in Paris,
devoted mainly
to the eating of hash
in hash-eating company

1887
Italian artist
Gaetano Previati
paints and exhibits
Women Smoking Hashish
a dire and desolate
interior landscape

1920
American author
Clark Ashton Smith
pens an epic poem
on eating hashish
that paints in words
what Gaetano failed
to paint on his canvas

1920 – 1940
Greece and China
ban hash smoking
while its cultivation
and use flourish in India
Bangladesh is known
as the Capital of Bhang

1980
Morocco becomes
the world's major
producer of hashish

1995
hashish is
sold openly
in Amsterdam
coffeehouses

today
this very instant

now I'm part
of this history
the trade and tradition
the expression
and I'm passing
it on to you

would you like a toke?

The Art Revisited

It was snowing dark smoke
and it was going to snow
dark smoke.

It was snowing
the story of tomorrow
and it was going to snow

the story of tomorrow.
It was going to snow
for a good time to come.

Falling steadily
and finding its way
into lives scattered

throughout the globe.
With black birds gliding
on the feverish winds

of that storm.
And the canvas
darker and more

detailed with age
than any artist
could fashion.

The Afterglow

Tendrils of illumination
Cling to my thoughts,
Trailing in my wake,
Puzzling to those
Whose paths I cross,
Those ever immersed
In the dull endurance
Of their daily tasks,
Without illusions,
Without perception
Of what lies beyond
The stolid borders
Of the everyday,
Insensate and
Unable to travel
In the domains
Of space and time
And consciousness.

The Mantle of Merlin

Earl Livings

Three days of fasting, three days of no water,
Three days of chanting, listening, fitful dreams,
Nothing to show for all my hopes and training,
After I pledged myself, this dark of moon,
To cavern silence, shape of dragon's claw.
In yellow cowhide, kneeling on worn rock,
I sway, start, shiver, pant, a cornered beast,
Striving to bear the wound that heals, reveals.
My eyelids close, snap open. Shadows reel
About the cave, half shapes of stags, wolves, birds,
White boar that killed all heroes, ashen snake
That slithers from my sight, to lure me out
And chase through forests thick with thorn and cries.

And then an alder tree, with skulls for fruit.
Their voices rise and fall, no wind, yet light
Pours out their eyes, with waves of splintered words,
A curse for those who sing without that gift
Of inspiration, salmon truths, dark tongue.
One skull shrieks, tosses wildly, drops between
My feet, splits open, swells into a cauldron
Bubbling with gushing, gaudy stars that whisper
Such power for the taking, such acclaim,
Halls ringing with applause and spinning gold,
Corrupt desires I wish I never had,
Who only sought the service wisdom sings.
Each star a lust I leave behind, each star
A warning spell, each star a veil of fear,
I plunge my face into their midst, wild fires

Scouring loose dreams, to leave me floating, hushed,
A sea-green borealis shimmering bold
Above, within, below, a mound of earth.
A sizzle-crack of lightning—mist shrinks, gathers,
Quickens the form of body, ragged robes,
My trembling limbs, my vision charged by breath.

Weighted dark, cold rock underfoot, no breeze,
A murmur, rhythm old as blood, from alcoves
Scattered about the cave, its phosphorescence
Not quick enough to pierce the seething gloom.
A spiral maze of footprints. Plumes of breath.
One figure, deerskin cloak, horned headdress, steps
From shadow. Points. And one in feathered cloak.
In hooded robes. In rags. With staff of bells.
With harp. With herbs and tomes. With possumed feet.
Ravens on shoulders. Beasts at heel. Soft growling.
Some naked but for sightless masks, tattoos.
Patterns of red and white. Incision scars.
All those who died before they died and live
After their death, divining word and deed.
Those who instruct, or yet may learn from me,
If I survive.

              They soon begin to sing
Or chant in tongues that stun my body's pulse.
Darkness about us swirls with livid scars,
With crashing arabesques, with sunburst chasms,
Omens embracing, trembling, shattering.
Death-flash of swords. War engines swooping. Cities
Flattened by whirling flame, then raised by towers.
Cavern reverberating, song and thunder.
Body reverberating, urge and awe.
Throat opens. Lips and tongue return that tune,
Embrace the rolling clouds of vision, summon
Bounty of herd and harvest, radiance

Of lovers, babes, delight of toil and hearth,
Tableaux of bliss that wavers instantly
Before the ravages of lack and sin.
The music wilts to jags of crimson mist
And, as the gloom refills the cave, the ground
Shudders and heaves, my body chills to silence.
I apprehend, amidst the gusting cold
And scent of lightning soon to strike, as body
Becomes as light as breath, it is my sight
That fails. I blink, squint, shake my head, and catch
Glimpses of shades and shapes, a sense of futures
Still coming into life. I lift my hand
And see its bones as twigs about to fall.
And deep within my darkness wisdom sighs,
A threshold held before me as the Muse.

Out of the mist she comes, like all first times.
Out of the hidden, gown of gold, her hair
A fiery nimbus, eyes like cauldron night,
Pale snake about her neck, her rowan voice
Speaking my name, my soul, ensnaring smile,
Her wreathed arms stretched towards me, palms upturned.
Though gloamed by fear of dust, by hollow hope,
By loss, the grief of broken lands and faith,
Distrust of song as cure and quest, I step
Onto the spiral path, and with each stride
Turning, turning tightly unto itself
The footprints poise my feet, my pulse, my voice.
Her secrets wait at the heart of all attention.
With first chill touch of pallid skin I scream,
Such searing pain, such splintering of thought,
Numbed being, smell of excrement, my own,
The blood, my own, the cleaving dark, my own,
My sense compelled to nothing but a sigh . . .

I shiver into hollow bones, lush feathers,
Hooked beak, honed talons. Launch into the gloom,

And climb, and bank, and fold my wings to swoop
Till air thrums, vision opens, breath as song,
Spread wings, enclose the light, the dark, the cave,
Enclose the churning past, the driving futures,
Soar as words spill from talon, wingtip, beak,
The roiling dark slowing to iridescence.

This moment reels. Once more a man, I bow
Before a standing stone, the white snake coiled
Beside the carved base, head upraised, its eyes
Regarding me. Nearby, the hazel tree
Above a pool infused with salmon light
Sways to the measure of my breath. I tread
In time around the rock, and know myself,
How lightning, storm and blast dismember me,
How inspiration gathers me at last,
Floods me with triple knowledge—what has been,
What happens now, what tales will always be.
The rock hums louder as I dance the gift
Of violet vision from the depths of silence.
I know the forms of song, those spells of change.
The speech of animal and star, why birds
Disguise their nests, will find them once again
In other seasons, after crossing seas.
The trails and songs of beast and fish, what fruits,
What seeds, what flower, will conceal, heal, kill.
The hare, the fawn, the wolf come to my bidding,
As do the lost, the dead, those not yet born.
I know the secrets under standing stones,
What light, what cries, arrive at every season.
How men will move across, above, beyond,
The roaring wheels, the wings with tongues of fire.
How dreams reveal and revel in our plight,
What news they bring, what griefs they wake us to.
Why many pardon massacre and rape,
Their frenzy speech, their eyes as dead as coins.

How best to curse those false in revelation,
Who stain the soul to plump the purse and skin.
Show me a blade of grass and I shall raise
A crystal cavern, crows above a battle,
The cries of mothers, daughters, wives, those tears
That fall like bodies breached by steel and flame,
Or sounds of trumpets triumphing the dead,
Or flames within all city walls, or feasts
Around a baubled pine, with gifts unwrapped,
The elegance of touch, a baby's laughter.

For lands and race will change as all things change,
A path of dragons, red through white through steel,
Each story serving river, tree, wind, star,
As an eye soon blinks, as a man soon nods awake,
The shadows of his shed pursuits now fled
From cave and mind, thresholds of apple blossom
Swirling around him while he stretches, stands,
Accepts the silver branch of bells, sings praise.

The Necromantic Wine

Wade German

Where wattled dragons redly gape, that guard
A cowled magician peering on the damned
Thro' vials wherein a splendid poison burns.
—GEORGE STERLING

In simultaneous ruin, all my dreams
Fall like a rack of fuming vapors raised
To semblance by a necromant, and leave
Spirit and sense unthinkably alone
Above a universe of shrouded stars.
—CLARK ASHTON SMITH

The blood-red sun begins its slow descent
Behind the distant, jagged line of peaks.
From this clear vantage on the flagstone roof
Where I have made a final hermitage
Of this abandoned tower in deep woods,
I watch those giant granite faces turn
From shades of grey to shade of cobalt blue;
And there above them, gliding on great wings,
I see the silver dragons in their flight
Returning to their eyries and high keeps.
The cool autumnal winds around me gust,
And now about me whirls a weirder breeze
Which whispers in my ear a rhyming rune—
And so an elemental speaks to me
Of her day's wanderings across the world,

And up into our planet's airless zones
That limited her flight to view the stars
Behind the vault of deep cerulean.
And now the wind grows wilder; she departs
To seek ethereal games with her own kind,
Amongst the changeling colors of the clouds
Aflame in twilight's final renderings.
The air grows cooler; so I step inside
And settle in beside the flame-fed hearth
To warm my bones, and smoke my briar pipe;
And lounging in narcotic quietude
I sip pale yellow wine and contemplate
The subtle incantations of the night.

But mortal issues rise to cloud my thought,
The same grey ghost that lately haunts the nights
Of this, my ancient age by sorcery
Sustained so many years beyond its span:
It seems I have grown weary of the world.
In youth, pursuit of wisdom was my quest,
And wonder, that bright star, had served as guide;
But somewhere in the passing centuries
Its incandescence dwindled; nearly dead,
That fulgent glow of wonder has gone out.
With what to stir the embers just a bit?
Despite the learning of three hundred years,
I never have held counsel with the dead;
I have but theories anyone might have
Of death's dimension and what lies beyond.
Ancestral imprecations on black arts
Have kept my line from straying to that gate;
But lately, I have pondered that old pact,
For there are other ways to gain the roads
Which dark magicians tread to seek strange truths;
I need not raise a corpse from its repose
By crude reanimation, or invoke

The wraiths who linger at unquiet graves;
I need not deal with ghouls in catacombs
Who sup on foul corruption in the crypt,
Nor need I bow to idols of dark gods.
Such methods—so impious and perverse!
There is a rarer magic, more refined
And suited to an acolyte of taste
Who would not risk an old familial curse . . .

I once discovered in a desert tomb
Strange hieroglyphs engraved upon a stone
That mentioned of a necromantic wine:
A darkling, ruby wine of philtered spells
Distilled in huge alembics of a dream
A demigod once dreamt who, dying, spilled
The poison in a glass canopic jar
Attendant demons slew each other for.
Another mention of the wine is here,
In this Lemurian scroll: it is described
As wine both sweet and bitter to the tongue,
With mystic operations on the mind
Inscribing arcane words of alchemy.
In one grimoire, the potion is compared
To green absinthe—pale opalescent drops
Evolving in the poet-prophet's brow
A third eye blazing like a demon star
That sees behind occulted nature's veil.
And one old libram notes the legend well,
But states the ruby potion is composed
Of substances abused by oracles:
The pollen of black loti thrice refined
And alkaloids from flowers of the moon,
Affording hypnagogic properties
On those who seek to see the dead in dreams.
And such I know of necromantic wine.
Who knows for sure what wisdom it imparts?

I have a bottle here; there is one thing
Betwixt this rare elixir's spell and me:
The cork.

        A darkness washes over me
Mere moments after sipping from the glass.
I shudder as a mist invades my mind,
The potion working like an anodyne.
My pulse throbs slowly, thudding as in sleep;
A sense of distance gathers in my head:
The chamber walls and ceiling now withdraw,
And all the candles glimmer distantly
Like witchlights in a black expanding pool.
I feel my body sink into the couch
And feel its fabric fray and then dissolve;
My atoms scatter as a thing destroyed.
Thus disembodied, and by wraith winds borne,
I am conveyed across the gulfs of night
And outer voids of undimensioned space
As swift hallucinations pass me by,
Successive strange horizons which unfold
Like tapestries, their imagery arrayed
In vast prismatic patterns which reveal
The surfaces of endless unknown worlds:
Strange vales and vistas, alien terrenes
With protean shores awash in pulsing hues,
The spans of all their suns and pendant moons.
But now the swarming throng of orbs disperse
And vanish out beyond my vision's reach
To merge with infinite immensities.
Now, in a region of black space, I see
A planet out of chaos newly formed:
Enormous storms that feed electric bale
Sweep red primordial skies with raving winds
As climates alternate in swift extremes.
Below the raging upper atmosphere,

Volcanoes bleed with endless lava flows,
And crimson rivers web a rifted main
Which quakes in primal night devoid of life.
And as the orb around its sun revolves,
Its smoking cauldron surface stills and cools,
And on it protoplasmic ichor gels:
Amoebic life-forms mindlessly evolve
And multiply at blind malignant rate—
The ancestors that spawn a fledgling race
Which treads across the dawns of centuries—
I see imperiums arise in time
And just as swiftly, witness their declines
By mode of nature or by work of man—
The cities lie collapsed in sunken seas
Or buried in abyssals of black sand;
The landscape quickly molders and decays;
The orb is now a planetary tomb
Where only subtle shadows faintly flit
Among the shrines and toppled monuments.
Again the vision fades. All sense deranged,
I hurtle through the interstellar deeps
And pass through regions of galactic cloud
Where I behold vast nurseries of stars
Which gleam like hellish rubies, xanthics, pearls
And fiery opals blazing into birth;
Then further, on accelerated course
Through unlit oceans of the outer dark
Until my flight decelerates in zones
Where Time's great gears have shuddered to a halt.
I stand upon the rim of the unknown.
Below me swirls a strange, phantasmal sea
In which converge wild raving cosmic streams
That gutter in fantastic cataracts
To feed the swirling whirlpool-gulfs below.
As if supplied by black ensorcelled lamps,
A weird dark radiance illumines all;

And from the gulf, huge shadow-things arise—
Twin ebon-bodied winged leviathans
With twisted limbs and long colossal claws.
They gather up dark matter in the gloom,
And from that substance, raise a massive gate
By thaumaturgic gestures. From its arch
Weird vortices of ectoplasm pour,
And in the gyrings, shapes of varied race
Rise up and multiply in manifold
Familiar shade—or take far stranger forms
Phantasmagoric, as in fever dream:
Of Titans, giants, gnomish folk, and imps,
And goblin beings, gargoyle-headed men;
And centaurs side-by-side with saurians,
Scale-tailed and crystal-eyed, in phantom ways;
And white arachnids, weirdly humanoid,
Which stride in spectral unison with things
Emerged from some mad god's menagerie—
Pale, luminescent algaes, many-eyed,
And faceless fungoid creatures, webbed and winged;
Odd floating orbs of psychic energy,
And other fabled forms innumerable,
Of otherworldly, unknown origins.
Now one thin wraith among the spectral throng—
Who is the only sample of his race—
Drifts forward as their sole ambassador,
And though he has no mouth with which to speak
I understand his language in my mind:

"We come in wonder, awe, and in our woe,
In death united and our knowledge pooled
(For what a shade has known all shades now know)
That one upon our portal is alive
Who treaded stars to seek our nebula:
Among our legions are the kings and queens,
The viziers, priests and wizards, generals

Of dynasties long dead, which ruled in realms
On planets orbiting the million suns
Your almagests and testaments assign
As white Subhel, and golden Azimech,
Blue Algol, and pale rose Aldebaran;
As orange Fomalhaut and Betelgeuse,
And Cabalatrab, red and emerald green;
And Genib, Iclil, Menkar, Deneb, Thuban,
Zedaron, Zaurak, Zubenelgenubi:
The alphas, betas, gammas in your charts
Which form the signs and symbols of the night,
The iconography of zodiacs.
The merfolk who once lived in cities spread
Beneath eternal vaults of lunar ice;
The globe-like beings of gas giant worlds
Who dwelled and drifted in pacific zones
Of atmosphere which like a cauldron brewed
Huge brooding storms that gathered gloom and churned
With centuries of crimson turbulence.
And others of an ever-changing shape
(For their true form is formlessness itself)
Who mimic those with whom they would converse;
And those who once inhabited no world
But flourished on the interstellar winds
Like motes of pollen borne upon the air;
And beings who once lived eternities
Perceived by others as a moment brief,
Like flashings of the subatomic sparks;
And others from an astral lineage
Who lived and died existences unseen
By those perceiving only matter's moulds;
And those enormous shadows over there,
Whose brows are furrowed by colossal glooms—
The ghostly pantheon of all our gods,
Whose avatars still haunt forgotten fanes
On worlds reclaimed by vast eternal night

In futile hope some acolyte of theirs
Might kindle at their altars some old faith.
Behold our ranks and files: the phantom host
That hails from sectors of the galaxy—
A spiral cluster, which, remotely viewed
From outer regions of the void, must seem
A mere amoeba in an ocean's mouth,
Whose own blind, futile gropings barely touch
The cold indifference of the universe."

The spirit legions all around me swirl
Like priests and ministers who would convene
An exorcism or some awful rite,
Discouraging my reeling mind with fear;
But speak instead the unimagined truths
Of lost religions, sciences and arts
Advanced by eon-ancient wizardries
They practiced once, and offer tutelage
In ways no sage or scholar could refuse . . .
But now their eldritch whisperings grow mute;
The vision fades, and rising from the fumes
That curl in primal chaos on my mind,
I hear a mausolean ocean's roar,
And in it, all the voices of the void
Break on emergent mist-enshrouded shores,
Disperse in hissing echoes, and recede
To voiceless shallows and the gloom-fed deeps.
All's silent now; again I am alone
Amid the vapors of a vanished dream.
The chamber walls and ceiling are restored;
My body has not moved, although I feel
A distance-ravaged traveler returned
To porch and portal in transfigured night;
And by the measure of an antique clock,
I know my voyage was a moment's dream
Evolved from out of only half a glass!

I have the answer to my query now—
I must imbibe much deeper. I would know
The mysteries those hosts of ghosts would teach;
Upon the threshold of their ebon gate
I shall convoke and summon forth a guide
To lead the way beyond. Then will I be
Enlightened for a strange eternity—
Or overwhelmed by horror in the end?
I quaff the strange elixir once again
And shudder as a mist invades my mind.
Familiars! Take from me these fleshy robes,
Then heap upon them these, my ancient bones—
This sorcerer departs!

Sandalwood

Michael Fantina

The scent of sandalwood across the sill
Now lures me to a dream by jinn prepared,
To realms high-walled and thick with mantic spells
Beyond which only fools and dreamers dare.
I see a wizard's brazier burning myrrh,
His mumbled spells form glyphs upon the air;
The brazier's fumes spool redly to the walls
As thronging shadows lengthen through the room.
He points one bony finger toward a door
And bids me go and lift the ancient latch.
And so it is I leave and find a lane.
Now here it is I take the narrow path
Past idols tilting in the icy rain.
The road rides higher toward the mythic hills
Whereon the crowns of castles whitely bulge.
I travel leas I've traveled here before
In other dreams both evil or benign,
And always it is Elwin that I seek,
Whose beauty and allure are as the stars
That burn with love and magic long unguessed,
Whose only thought is for me and my love,
As in each dream I, the great knight errant,
Have won her from a host of evil things
That slither, slide or shamble with a will
To harm her or destroy me ere my quest
Is done and I have won her true and free.
I've slain the dragon and the hydra there,
Within my dreams, and end in Elwin's arms,
The beauty of a Siren, so demure

And like some goddess in her winsome ways.
My dream has well begun. I see the cairns
Wherefrom slim sallow ghosts will rise to warn
That I turn back ere the night grow too dark
And things that no man ought to see will come.
I dismiss these spirits for well I know
That I will best these beasts of sanguine dream,
Win to the lips of my beloved girl,
Drown in her hair, swoon at her whispered words.
The wind is up and stiff, I know a chill,
A chill perhaps I have not known till now
In all my panoply of darkling dreams.
I mark the vagrant comets overhead,
I shrug, and move into this biting wind.
I wrap my coat more tightly and my hand
Holds fast the staff that I have now procured,
A crosier, gift from errant hippogriffs
Who bid me well and take their leave withal.
My quest, for such it is, will take me to
Fay Elwin's castle where the lightning strikes,
Whose minarets and keeps frown darkly down.
An ancient arcane curse has walled her in,
She both princess and sorceress, enthralled
By spells outré and transmundane, while I
Will breach her walls and take here hence, away.
I come before her monstrous door and strike
My iron fist against the iron gate,
The echoes echo down the valley far.
I shout the name of fairest Elwin there.
And then the great gate shudders, falls. I'm in!
I sweep away a host of seneschals,
They cringe away or die by my strong staff.
I'm led by spirits to a winding stair,
And up and up into those dizzy heights
I climb so breathless for my Elwin fair
I dream her there supine, her golden hair.

A one-eyed troll now guards her door alone,
His iron cudgel and his giant hands
Prepare to smash me to such smithereens.
Once more I am too quick, and know the dream;
I fling him headlong down the winding stair,
His wailings echo loudly through the halls.
I throw the bolt upon my Elwin's door
And enter in, excited, nervous, yet
In control, quite supreme within my dream,
The scent of sandalwood so strong, and now
I see her there, awake, upon a chair.
"My lover," says she, and smiles, so oddly,
No smile of love but one of mockery!
"Come forth, for this is *my* dream now and you
My willing thrall, my will you'll work with zest,
My every wish a swift command to you!"
No more my dream but hers! She is supreme,
I'm but her pet to do with as she will!
Her eyes so like blue pools in mountain vasts,
Her bare and narrow waist there girded round
With tiny chains, fine strand on golden strand,
Her navel holds a ruby like some eye,
So huge like to a Cyclops' orb it glares.
Her hair falls long and golden in thick curls,
One gorgeous leg poked through her skirt's long slit.
I know she speaks the truth I am her slave,
This dream is hers, no more a dream at all,
A nightmare now held under her harsh rule.
Away we flee to realms of ice and snow,
Where witches rule and elementals war,
Where comets light a landscape full of gloom,
A moon rules over all—there is no sun.
We stand knee deep in snow, she turns and smiles,
"I grow so bored of love and dreams, and you,
My hapless human dreamer, yet there's sport:
I'll make of thee a pawn to work my will,

To quench my loneliness and black ennui.
Here is a sword, O dreamer. From the ice
Will come a monster, slay him here for me!"
Before my frightened eyes uplifts a thing
With large serrated teeth in each huge head,
Its two legs squat and thick, with massive hands,
As agile as an ape, its eyes black bale.
I quail before the creature and I flee,
I fling my sword aside, yet am pursued,
Not by the thing, but worse, by Elwin then!
She bellows and she swears to see me dead.
In vain I try to wake, but no, she comes
The closer as I hurry through the drifts.
I mark a cave's mouth there upon my right
And into it I enter and I hide.
Elwin enters, as lovely as the dawn,
Her eyes alive yet full of wickedness.
She speaks: "Ah, lover, all mine own! Come out!
And I will show thee love so far unknown
That you will always stay with me, in dream!"
In terror now I long to wake, to leave,
To flee my dreams and never to return.
And now I watch as Elwin puts her hands
Above her pretty head, her body shakes,
Her agile form becomes a cobra there!
She slithers near to find me in the dark!
I bolt now, run back deep into that cave
Whose icy terrors few might ever guess.
Deep, deep into that cave I flee in fear,
Down, down, I run with her in hot pursuit,
Her laughter echoes from these walls of gneiss.
I trip and fall on skulls of ancient beasts,
Some single eyed, some with tri horns, all pale
Who grin at me with massive mockery.
Yet still I run, and still I hear her steps.
I move through tangled webs so wan and thick,

Disturbing black arachnids large as rats,
And still dark Elwin gains upon me quick.
I feel a potent thing now grasp my leg
There, just above the ankle, and she laughs.
My lamp appears beside me in my room,
She stands, a girl again, sweet Elwin now,
A dagger in her hand. "Die my lover!"
I move my arm and hit the nearby lamp
And tumble there amid her laughing voice.
I pray and wait then for that blow to fall,
As sandalwood fills all my lungs, and I
Now sigh for fear, the very crack of doom
Upon me in the dark, and yet the cold
Is gone. I lay alone, within my bed,
Where scents of sandalwood grow strong and thick.
My Elwin gone, she but the ghost of dream,
And all those realms mere vapor in the air,
Though still, what of the sandalwood?

LUCUBRATION

Kyla Lee Ward

A composition that smells of the lamp . . .

A most perplexing paradox, to write
so late at night with such a light as this
banishing darkness from the magic ring;
but day thoughts are not night thoughts, so this hiss
of warming gas and stink of burning dust
must be the very consecrated 'cense,
the silver inlay and the mighty name
that lets me press the demon with a slim,
unsure advantage. Is this but his play,
awaiting chance to reft my soul away?
Still I would summon darkness! That each door
and window, underside and crack extend
into the furthest reaches of the 'byss,
all that has been, may come to be and is.

It seems that I have stalked tenebral paths,
leaves pressing thick and bends obscuring sight.
Only the scent of cereus and of rose
to mark the way: the velvet brush of moss
and kiss of web assurance of retreat.
This is my victim's garden, well beyond
the outer wall yet feel I still no fear—
a mask and blade ensure my welcome here.
To move so swift, so silently and sure!
No thorn ensnares, no twig betrays my step;
the grace of darkness speeds me to our tryst

in empty passage, else his very bed.
No matter whose the coin that bought this death,
of all that call this rotting city home,
there shall be heat, a muffled scream and blood
sweet on my lips, for I am nothing more
than wolf or 'pard, or any other bane
they seek to keep beyond the palisade.
Close now, all unaware my prey awaits,
a turn, and through the shivering of fronds
and shattering of water, there is light.

I see them through the arches where they dance
gilded by harpsichord and violin,
silvered by laughter and flirtatious glance;
none look without. To them the world is all
a crystal ball that catches candlelight
and sends motes spinning, hand in tender hand.
In peacock satin, broidered and bejewelled,
their faces masked as panther, wolf or skull
and he, 'tis surely he who thus affects
assassin's black, a blatant mockery.
For I must never stray within the light,
the legendry of my cruel craft maintain,
no partner take, save in the dance of death;
this code enshrined, yet how may I refrain?
Why do I stalk these sightless alley-ways
but for a glimpse of life's rose aureole?
Why take their coin and venture in their sphere
if not to touch all that I was denied
when fate cast me upon the darker shore?
I dance upon a dagger every night!
Why should I not dance here amongst the crush
of hip and shoulder, lacing lip and thigh?
How could they tell my visage from their own?
So entered I, and soon the answer learned,
such answer as permits the gods to laugh,

such grace I had as lay beyond the dance;
where others faltered, I descried the tune,
nor caught a foot, nor brushed a careless hand
in weaving ever closer to my prey:
oh laughing gods, unravel how it may
be grace itself that gave myself away?
Questions assail me, laughter closes ranks
and hems me in. I can see nothing past
the light but light permits them to discern
the deed undone and mark, it is for that
I die, beyond a hundred crimson crimes.
Now as my judges bring the rod and flame
I cry for darkness, prodigal in shame
and only dread the work that they now do
will leave life's final shackle on my wrist.
All I have been brought screaming down to this.

Thus I retreat, and seek throughout the vast
for certain sanctum and yet deeper dark.
It seems I have walked colonnades aside
of hornéd heads with women's breasts, and known
a man's face rise above a lion's paunch,
serpents and scarabae with human hands
and phalluses: I say, not all were stone.
Across the sands disguising all above,
devoutly, pilgrims trace an ancient path
from out the lesser shadows of the night,
down cunning stairs that lead to us below.
Echoing vaults as chill and black as death,
where barques of granite draw their cargo nigh,
and stone papyrus hold the heavens high.
Our oracle brings men with azure beards
and layered robes, redheads with pallid skin,
bronze men and black: with offerings of wood,
of iron and silver, ivory and salt,
the oil of whales and one thing more, for here

no circle holds the demon kind at bay.
The supplicants pass columns in the murk,
offering bodies to a winged embrace
and throats to kisses bringing such sweet pains
as only teeth permit and blood contains.
They come for knowledge: knowledge they shall find
in scented smoke that frees the untrained mind,
murmured by shades and hissed by coiling fiends,
and whispered from the lips of blessed things
that pass above them, stirring fragrant wind.
Whate'er they seek, be it the fate of kings,
the course of wars, felicity of brides,
or cure of plagues: the answer here abides.
Yet none of these shall ever find the lake
where lotus blossoms raise their scented heads
above black waters, warm and thick as blood.
None shall approach the greatest mystery,
the beating stone, the nigresence within
the inmost shrine, where only priests may go.
Only the chosen: all these paths are mine.

Yet still there is a thing that troubles me:
dogging my step and stinging in my eye.
Not ghost nor genius, yet it flits along
the colonnade, a disc both flat and fleet.
I fail to catch it: then as I pursue
I recognise intrusion from above.
Within the ceiling gleams a single hole
and through that hole there lies a shining world
of tincts known only through their dying fall,
of sounds and scents, of trees and fields of grain,
loud rivers flowing underneath a sky
where rides a god that shows himself to all.
I am of those who see without their eyes,
kept from the world to force the subtle sense:
and though I now imagine sight and sound

and quake within, I know my duty well.
With sleeve across my face, a fragile mask,
I seek out proper servants for the task
of sealing up this breach of sanctity.
Had I but looked, perhaps I would have seen
the doom that was foretold us long ago
approaching now. Perhaps I could have been
the saviour of some shard, and wrought a fate
somehow less cruel, and kept our memory
intact for all the ages yet to come.
The dust sifts down, and yet I see it not.
The columns creak, yet I walk on until,
with booming shriek and rushing tide of sand,
the stony heavens split apart in flame
and men descend who seek no wisdom here.
The spirits shriek and yet they hear them not.
The monsters writhe but they can only see
statues adorned with gold and many gems
run molten in the furnace of the day.
All I might be is given to decay.

Once more, once more I flee into the dark
and, shrieking, seek out such a potent form
as may defend the fragments of my soul.
No weakness now, of either love or fear.
This blade I bear is forged of ancient shards.
This mask I wear was stolen from a tomb.
Armour I carry, and an evil name,
that Christian priests may use to conjure hate.
I am the wolf's right hand, the raven's throat,
who rises from the forests of the north
where endless run the trees whose blackened trunks
and gloomy branches scarify the sun
and men mistake the days for nights until
dread madness seize them, else they chance on me.
I come now, from the necromantic groves

as crest of storm devouring the bright day
and all the slaves in fields and cities quake:
their devil curse, that loosed me on the world.
Their devil, ha! They'd better blame their God.
Dreadful the sacrifice that has allowed
my conjuring, and it was not my hand
which spilled the blood across the altar stone.
In my ranks heretics and rebels crawl
who lost their lives upon the block or wheel,
and lovers slain for loving past the bounds,
proud pagans Roma never held in thrall,
the charcoal husks of witches, and yet more:
monsters and prodigies of form possessed
to make of man's perfection but a jest.
All share my hunger to extinguish light
and in this final victory to rest.

"O hear me now, you captives of the hour!
Yes, hear me now, you blind and shackled fools!
Prepare to welcome all you have despised.
The cross shall not avail you, nor the scourge.
Embrace us, love the grand catastrophe,
for by this turn or death, you shall be free!"
Walls crumble hissing into ancient sand
at my bare touch; the sputtering cannon dies
at merest glance, and where I tread the grass
is blackened and the earth gapes wide and births
yet more abominations at my call.
So die the knights, their armour but a glove
to animated tendrils of the dark;
so die the priests, their chastity consumed
by laughing nightmares, bringing bliss with fangs;
so die the mothers, shrieking out their right
to mercy to rough things of stone and clay
that never knew the pressure of the womb.
So die they all, and every child who cries.

The beacon fires, lanterns of the watch,
the flambeaux that illumine all the State,
the great cathedral seeming to contain
within a rose of glass the very sun,
now one by one, they die. All die and night
comes sweeping through the city like the tide,
the final flood that never shall recede.
One light remains within the highest tower,
one light alone, and this both faint and far,
but I'll not leave my vengeance incomplete.
My best companions running at my heels
and screeching in my wake, I take the stair
that circles ever upwards, ever on.
The light's a grin that mocks my every step!
The light's an eye that sees my deepest pain.
The light: the light is no more than a lamp
that casts a circle round a desk and chair,
and adumbrates the figure seated there
with pen in hand. And yet somehow in this
I see creation's round and all that is.

If I wrote in the darkness, the result
would be such scratchings and strange hieroglyphs
that any passing eye would think no more
than random marks, the proper work of flies
or imbeciles. If such the demon is,
my own self come to steel me to this task,
the greater soul of which I am the part
that weeps, my duty nonetheless is clear:
complete the work that is my purpose here,
as I return to darkness, so to light,
without whose hissing breath, I cannot write.

SELECT BIBLIOGRAPHY

"The Hashish-Eater; or, The Apocalypse of Evil." 1920. 581 lines. In *Ebony and Crystal* by Clark Ashton Smith (Auburn, CA: Auburn Journal, 1922); rpt in *Selected Poems* by Clark Ashton Smith (Sauk City, WI: Arkham House, 1971); rpt in *Clark Ashton Smith: The Complete Poetry and Translations*, ed. by S. T. Joshi and David E. Schultz (NY: Hippocampus Press, 2007–08; rpt in paperback 2012).

"Lucubration." 2012. 231 lines.

"The Mantle of Merlin." 2012. 150 lines.

"Memoria: A Fragment from the Book of Wyvern." 1976. 150 lines. In *Spores from Sharnoth and Other Madnesses* by Leigh Blackmore (Sydney: P'rea Press, 2008–).

"The Necromantic Wine." 2012. 275 lines.

"Sandalwood." 2012. 155 lines.

"Thirteen Ways of Looking At and Through Hashish." 2011. 510 lines.

"A Trip to the Hypnotist." 2011. 100 lines. In *Intimations of Unreality: Weird Fiction and Poetry* by Alan Gullette (NY: Hippocampus Press, 2012).

"Visions of Golconda." 1959. 132 lines. In *Savage Menace and Other Poems of Horror* by Richard L. Tierney (Sydney: P'rea Press, 2010–).

"A Wine of Wizardry." 1904. 210 lines. In *A Wine of Wizardry and Other Poems* by George Sterling (San Francisco: A. M. Robertson, 1909); rpt in *The Thirst of Satan: Poems of Fantasy and Terror* by George Sterling (NY: Hippocampus Press, 2003); rpt in *George Sterling: The Complete Poetry*, ed. by S. T. Joshi and David E. Schultz (NY: Hippocampus Press, 2012).

About the Contributors

LEIGH BLACKMORE (BCA Writing, Hons) is no stranger to wizardry, having been a dedicated ceremonial magician for over 25 years. As President of the AHWA (2010–2011), he edited *Midnight Echo 5*. He runs an editorial business, Proof Perfect, and edits SSWFT Amateur Press Association. A widely-published critic, editor, poet and story writer, Leigh has twice been nominated for the Australian Ditmar awards (for fiction and criticism). Recent work appears in *Studies in the Fantastic* and in *Weird Fiction Review*. His verse collection *Spores from Sharnoth and Other Madnesses* (P'rea Press, 2008–) garnered extensive acclaim. For more information see: http://www.australianhorror.com/member_pages.php?page=86

BRUCE BOSTON lives in Ocala, Florida, with his wife, writer-artist Marge Simon, and the ghosts of two cats. He is the author of fifty books, including the novels *The Gardener's Tale* and *Stained Glass Rain*. His writing has appeared in countless publications, most visibly in *Asimov's SF Magazine*, *Amazing Stories*, *Weird Tales*, *Year's Best Fantasy and Horror*, and *The Nebula Awards Showcase*. One of the leading genre poets for more than a quarter century, Boston has won the Bram Stoker Award for Poetry, the Asimov's Readers Award, and the Rhysling Award (SFPA), each a record number of times. www.bruceboston.com

MICHAEL FANTINA has been writing fantasy poetry for more than four decades. His poetry has been published in North America, Australia and in the United Kingdom. He has appeared in such publications as *The Romantic Quarterly*, *The Penwood Review*, *The Lyric*, *The New Formalist*, *Midnight Echo*, every number of *Candelabrum Poetry Magazine* (1996–2007) save one, and numerous other publications. His fourth chapbook of verse, *Ghosts of the Sand*, will appear from Rainfall Books sometime in 2013. Michael has also sold horror fiction to magazines in the US and Japan.

WADE GERMAN's poems have appeared internationally in numerous journals and anthologies, including *Dark Horizons*, *Dreams and Nightmares*, *Fungi*, *Midnight Echo*, *Mythic Delirium*, *Nameless*, *Phantom Drift*, *Space and Time*, *Star*Line*, *Strange Sorcery*, and *Weird Fiction Review*. His work has been nominated for the Pushcart and Rhysling awards, and has received honorable mentions in Ellen Datlow's Best Horror of the Year anthologies (volumes II and III). In 2012 he guest edited weird verse issue 6 of *Eye to the Telescope*. He maintains a strong interest in weird literature, art, and poetics.

ALAN GULLETTE began this incarnation in 1956, as a Cancerian, and Floridian. He was early drawn to mythology and weird literature. At 14 he began writing pastiches of Poe, Lovecraft, Dunsany and Smith. At 16 he published his own fanzine, *Ambrosia*. By college his

interests included mysticism, philosophy, chess and music, with readings in Surrealism and Absurdism. His poetry books include *From a Safe Distance*, *49 Pieces*, and *Twenty-seven Liqueurs*. An omnibus collection of fiction and verse, *Intimations of Unreality* was published by Hippocampus Press in 2012. Now Californian, he is married to Julie Hodge; they have one cat.

S. T. JOSHI is the author of *The Weird Tale* (1990), *The Modern Weird Tale* (2001), *I Am Providence: The Life and Times of H. P. Lovecraft* (2010), and *Unutterable Horror: A History of Supernatural Fiction* (2012). He has prepared annotated editions of the fiction of H. P. Lovecraft, Ambrose Bierce, Arthur Machen, and other authors, and has edited the collected poetry of Clark Ashton Smith and George Sterling. His slim treatise, *Emperors of Dream: Some Notes on Weird Poetry*, was published by P'rea Press in 2008. He is the editor of the *Weird Fiction Review* and the *Lovecraft Annual*.

EARL LIVINGS was born in Melbourne, Australia. He studied mathematics, played guitar in a garage band and gained a black belt in kung fu. Earl has had poetry and fiction published in journals and anthologies around Australia and also in Britain, Canada, the USA, and Germany. His first book of poetry, *Further Than Night* (Bystander Press), was published in 2000.

He teaches professional writing and editing and is the editor of the poetry journal *Divan*. Earl lives in Melbourne with his wife and two cats and is currently working on a novel and his next poetry collection.

CHARLES "DANNY" LOVECRAFT is an editor, publisher, writer, and versifier. He began his apprenticeship with literature and his love affair with poetry at an early age by the special influence of H. P. Lovecraft. More than one hundred of his poems have been published in magazines and anthologies. He has also authored a bibliographic checklist of Richard L. Tierney's work and edited twenty-nine books. In 2007 he established P'rea Press to publish international weird and fantastic poetry and non-fiction. Charles researches classic Australian supernatural poetry and promotes reading, writing and performance of fantasy poetry by conducting panels and workshops at conventions.

GAVIN L. O'KEEFE has been creating artwork for a range of literary genres for over 25 years, and his illustrations have appeared in books and magazines worldwide. His illustrations for Lewis Carroll's "nonsense" books were published by Ramble House as *The Alice Books* (2011) and *A Snark Selection* (2004), and his illustrations for the time travel stories of H. G. Wells and Richard A. Lupoff appeared in *The Book of Time* (Surinam Turtle Press, 2011). Gavin's work includes many book cover designs for crime, horror and science fiction novels and collections. Born in Australia, he now lives in Maine, USA.

The artwork of DAVID SCHEMBRI has appeared in *Midnight Echo*, *Andromeda Spaceways*, *Black*, and on the AHWA website. On the literary side, his first novelette "The Unforgiving Court," appeared in the Chaosium anthology *Undead and Unbound*. His "The Black Father of the Night," was published in the Horror World Anthology, *Eulogies II*. A graphic collection, *Unearthly Fables*, edited by Paula B. of *The Writing Show*, was published in 2014. The book features his prose fiction and macabre artwork. David lives in Australia with his lovely wife and children.

For more than two decades, DAVID E. SCHULTZ has co-edited (with S. T. Joshi) numerous editions of the writings of H. P. Lovecraft, Ambrose Bierce, George Sterling, Clark Ashton Smith, and others. His edition Lovecraft's *Fungi from Yuggoth* was published in 2016. He is chief book designer for Hippocampus Press and P'rea Press, and has designed books for Dark Renaissance Books, The Edwin Mellen Press, Mythos Books, Scarecrow Press.

CLARK ASHTON SMITH (1893–1961) attained early celebrity with the poetry volumes *The Star-Treader and Other Poems* (1912), *Odes and Sonnets* (1918), *Ebony and Crystal* (1922), and *Sandalwood* (1925). Later he wrote more than 130 tales of fantasy and horror, collected in such volumes as *Out of Space and Time* (1942) and *Lost Worlds* (1944). Smith lived most of his life in Auburn, California, and was a close friend of George Sterling, H. P. Lovecraft, and August Derleth. His *Complete Poetry and Translations* appeared in 3 volumes in 2007–08.

GEORGE STERLING (1869–1926) was born in Sag Harbor, Long Island, but moved to California in the 1890s, where he remained until his death. He gained local celebrity by such poetry volumes as *The Testimony of the Suns* (1903), *A Wine of Wizardry* (1909), *The House of Orchids* (1911), and *Sails and Mirage* (1921); his *Selected Poems* appeared in 1923. He became the informal "king of Bohemia" and was close friends with such figures as Jack London, Clark Ashton Smith, Theodore Dreiser, and H. L. Mencken. His *Complete Poetry*, including his verse dramas, appeared in 3 volumes in 2012.

For more than half a century RICHARD L. TIERNEY has composed uncompromisingly delicious dark poetry and prose. Born whilst H. P. Lovecraft was still alive, he has been one of the foremost Lovecraftian poets of modern times. His versatility has also led him to write poetry reminiscent of Robert E. Howard and Clark Ashton Smith. Like the barbarian flames in sacked cities of antiquity, in everything Tierney writes a fierce originality burns through the lines. His poetry volumes include *Dreams and Damnations* (1975), *The Doom Prophet* (1976), *Collected Poems* (1981), *The Blob That Gobbled Abdul* (2002), and *Savage Menace* (2010).

KYLA LEE WARD is a Sydney-based creative who works in many modes. Her latest release is *The Land of Bad Dreams*, a collection of dark and fantastic poetry. Her novel *Prismatic* (co-authored as Edwina Grey) won an Aurealis Award for Best Horror. Her short fiction has appeared on *Gothic.net* and in the *Macabre* and *New Hero* anthologies, amongst others. Role-playing games, short films and plays, art—if you can paint it black she probably has. A practicing occultist, she likes raptors, swordplay and the Hellfire Club. To see some very strange things, try http://www.tabula-rasa.info

Avatars of Wizardry by George Sterling, Clark Ashton Smith, et al.
(November 2012; rpt, February 2016)
ISBN: 978-0-9804625-8-6 (paperback) $14.50AU
ISBN: 978-0-9804625-9-3 (ebook) $10AU

OTHER PUBLICATIONS FROM P'REA PRESS

Richard L. Tierney: A Bibliographical Checklist by Charles Lovecraft.
(February 2008)
ISBN: 978-0-9804625-0-0 (paperback) $6AU

Spores from Sharnoth and Other Madnesses by Leigh Blackmore.
(September 2008; rev. rpt, August 2010, May 2013, February 2016)
ISBN: 978-0-9804625-2-4 (paperback) $14.50AU

Emperors of Dreams: Some Notes on Weird Poetry by S. T. Joshi.
(November 2008)
ISBN: 978-0-9804625-3-1 (paperback) $14AU
ISBN: 978-0-9804625-4-8 (hardcover, out of print)

Savage Menace and Other Poems of Horror by Richard L. Tierney.
(April 2010)
ISBN: 978-0-9804625-5-5 (illustrated numbered hardcover) $30AU
ISBN: 978-0-9804625-6-2 (illustrated ebook) $10AU

The Land of Bad Dreams by Kyla Lee Ward.
(September 2011; rpt, February 2016)
ISBN: 978-0-9804625-7-9 (illustrated paperback) $14.50AU

Dark Energies by Ann K. Schwader.
(August 2015; rpt, August 2015; February 2016)
ISBN: 978-0-9943901-0-3 (illustrated hardcover) $26AU
ISBN: 978-0-9804625-1-7 (illustrated paperback) $14.50AU
ISBN: 978-0-9943901-1-0 (illustrated ebook) $10AU

P'REA PRESS

Publishes weird and fantastic poetry and non-fiction
c/–34 Osborne Road, Lane Cove NSW Australia 2066
Website: www.preapress.com
Email: DannyL58@hotmail.com

www.ingramcontent.com/pod-product-compliance
Lightning Source LLC
Chambersburg PA
CBHW031256290426
44109CB00012B/612